STARK EFFECT IN A HYDROGENIC ATOM OR ION

Treated by the Phase-Integral Method with Adjoined Papers by A Hökback and P O Fröman

STARK EFFECT IN A HYDROGENIC ATOM OR ION

Treated by the Phase-Integral Method with Adjoined Papers by A Hökback and P O Fröman

Nanny Fröman
Per Olof Fröman
University of Uppsala, Sweden

Imperial College Press

Published by

Imperial College Press
57 Shelton Street
Covent Garden
London WC2H 9HE

Distributed by

World Scientific Publishing Co. Pte. Ltd.
5 Toh Tuck Link, Singapore 596224
USA office: 27 Warren Street, Suite 401-402, Hackensack, NJ 07601
UK office: 57 Shelton Street, Covent Garden, London WC2H 9HE

British Library Cataloguing-in-Publication Data
A catalogue record for this book is available from the British Library.

STARK EFFECT IN A HYDROGENIC ATOM OR ION
Treated by the Phase-Integral Method with Adjoined Papers by A Hökback and P O Fröman

ISBN-13 978-1-86094-924-1
ISBN-10 1-86094-924-X

Typeset by Stallion Press
Email: enquiries@stallionpress.com

Printed in Singapore by World Scientific Printers

Preface

The purpose of the present book is to use the phase-integral approximation of arbitrary order generated from an unspecified base function, which is described in Chapter 4, in a systematic treatment of the Stark effect for a hydrogenic atom or ion in a homogeneous electric field. Previously the Carlini[1] (JWKB) approximation has often been used to treat this problem, and there have many times appeared discrepancies between results obtained by the use of that approximation and accurate numerical results. As has been pointed out in particular by Farrelly and Reinhardt (1983) the reason for this is in general not that the approximation method is inadequate, but that it has often been used incorrectly. When it is used in an appropriate way, the discrepancies disappear, and one often obtains highly accurate energies for the Stark problem already in the first-order approximation. This conclusion applies, even to a larger extent, to the approach based on the still more efficient phase-integral approximation of arbitrary order generated from an unspecified base function, which for the Stark effect in hydrogenic atoms or ions yields analytical formulas expressed in terms of complete elliptic integrals, which can be evaluated efficiently by means of standard computer programs.

In Chapter 8 a large number of phase-integral results are compared to results obtained by other methods. Of the 198 different states discussed there, which correspond either to different quantum numbers or to the same quantum numbers but different electric field

[1] As regards the motivation for the name Carlini approximation we refer to Fröman and Fröman (1985) or to Chapter 1 in Fröman and Fröman (2002).

strengths, the phase-integral energy values are judged to be at least as accurate as the results obtained by other methods in more than half of the cases. It may also be mentioned that, compared to the results obtained by other methods, the phase-integral energy values can contain up to six more digits.

We would like to thank Professor R. J. Damburg and Professor V. V. Kolosov for valuable correspondence and Professor M. Lakshmanan for letting us share his experience concerning elliptic functions and elliptic integrals. The results presented in Chapter 7 and Chapter 8 could not have been obtained without Research Engineer Anders Hökback's first-class work. We are also indebted to him for having drawn the figure in Chapter 1 with the use of a computer program. During decades we have had the privilege of having close contact with Professor Ulf Uhlhorn and have profited very much from his great scientific knowledge. Concerning the present book we have had many useful discussions with him. In particular he has helped us to draw the figures in Chapter 5 with the use of a computer program.

Nanny Fröman
Per Olof Fröman

Contents

Chapter 1

Introduction

Since the Schrödinger equation for a hydrogenic atom or ion in a homogeneous electric field is separable in parabolic coordinates, the system is more easily accessible to an accurate theoretical treatment than other atoms or ions in electric fields. Furthermore, a Rydberg state of any atom or ion, i.e., a state in which one electron is in a highly excited state and the nucleus is shielded by the core of the other electrons, resembles a hydrogenic state in the sense that a single electron moves far away from an ionic core and does not penetrate into the core unless the magnetic quantum number m is small enough. Such an atom or ion is thus approximately hydrogenic. In the interstellar space there occur very highly excited Rydberg atoms, with values of the principal quantum number of the valence electron of the order of magnitude of one hundred. They are therefore closely hydrogenic and may be exposed to strong electric fields. With the ever increasing accuracy and sofistication of experiments the need for accurate analytical methods of analysis will increase, and the treatment of the Stark effect in hydrogenic atoms or ions may serve as a model problem for the treatment of Rydberg states. Thus, apart from the intrinsic interest of its own, the Stark effect problem for a hydrogenic atom or ion plays the role of a model problem, from which one can obtain information about the properties of Rydberg states. For comprehensive reviews of the properties of Rydberg states, see Gallagher (1988, 1994).

The quasistationary nature of the Stark resonances, due to the fact that the energy eigenvalue spectrum of one of the coupled differential equations, resulting from the separation of the Schrödinger equation in parabolic coordinates, is continuous, with wave functions extending to infinity, makes the hydrogenic Stark effect problem intricate in spite of its seeming simplicity.

Brief review of different aspects studied and various methods used

Precise experimental results on Stark levels in atomic hydrogen have been reported by many authors, and a great number of theoretical papers have appeared in which different methods are used for the study of the Stark resonances. Semi-classical methods are adequate for highly excited states and have been used by several authors. The Stark effect for levels well below the top of the barrier was treated with the aid of Carlini (JWKB) technique, although rather crudely, already in the early days of quantum mechanics. Later there appeared improved treatments of that kind, in which also levels near the top of the barrier were considered. Another important method for determination of the positions of the Stark levels is the Rayleigh–Schrödinger perturbation theory, but it is not applicable in low orders for highly excited states and strong fields. However, perturbation theory of the Stark effect in atomic hydrogen has been made tractable to arbitrarily high orders by a restatement of the perturbation theory formulas that allows the perturbation series to be obtained from recursive relations run on a computer. The perturbation series is not convergent but asymptotic, and Borel summation together with the use of Padé approximants greatly accelerates the approach towards accurate energy values and is an efficient tool for obtaining accurate results for the Stark effect. It will, however, not be discussed in this book, since we restrict ourselves to the use of the phase-integral approximation generated from an appropriately chosen base function. This approach, which is capable of yielding *explicit analytical* formulas, is for the first time applied to the Stark effect in a systematic way in this book.

There occur in the literature different ways of defining the positions and the half-widths of the resonance levels. One finds definitions based on considerations of the probability amplitude, or based on the rapid variation of the phase shift with energy, as well as definitions relating the real part of a complex energy eigenvalue to the position of the Stark level and its imaginary part to the width of the level. Different aspects of the last mentioned approach, i.e., the use of complex energy eigenvalues, and comparisons with different formulas for the half-width at low fields can be found in Yamabe, Tachibana and Silverstone (1977). For narrow levels the differing definitions yield essentially the same numerical results. However, for broad levels slightly below the top of the potential barrier and for autoionizing levels above the top of the barrier the differing definitions yield appreciable differences in the results. The broad levels are not of Lorentzian shape but are highly asymmetric, and hence the concept of half-width loses to some extent its precise meaning. As a consequence of this fact it is obvious that the methods based on complex energies are inadequate for broad levels.

The advent of tunable lasers created a radically new situation as to the possibility for selective excitation of high Rydberg states and for making precise measurements on their properties. Highly excited atoms are very sensitive to external fields, and currently used field ionization methods are very powerful for detecting Rydberg states. As a consequence of these circumstances, such an old problem as the Stark effect in atomic hydrogen attracted a renewed interest.

Brief account of the background of this book

Papers concerning the Stark effect of a hydrogenic atom in a homogeneous electric field appeared already in the early days of quantum mechanics. On the basis of the matrix mechanics invented by Heisenberg (1925), Born and Jordan (1925), Dirac (1925), and Born, Heisenberg and Jordan (1926), Pauli (1926) obtained for the spectrum of the hydrogen atom and for the Stark effect of that atom results that agreed with experimental data. In connection with his development of wave mechanics Schrödinger (1926) made an application to the Stark effect in atomic hydrogen. He separated the time-independent Schrödinger equation for the problem in question in parabolic coordinates and used first-order perturbation theory to treat the two resulting ordinary differential equations.

In the paper where Wentzel (1926) presented his rediscovery of the Carlini (JWKB) approximation, he applied this approximation to the treatment of the two ordinary differential equations just mentioned. Waller (1926) treated instead these two ordinary differential equations by expressing their solutions with the use of series expansions in powers of the field strength. By successive approximations he obtained a second-order formula for the energy levels of a hydrogen-like ion in a homogeneous electric field. Only slightly later Epstein (1926) also presented a theory for the Stark effect in a hydrogen-like ion, based on the time-independent Schrödinger equation, which he, after separation in parabolic coordinates, treated by successive approximations and obtained results up to the second order in the electric field strength. Van Vleck (1926) used the formula for the energy levels of a hydrogen atom in an electric field, obtained independently by Waller and Epstein, to calculate the dielectric constant of atomic hydrogen. The Stark effect in hydrogenic atoms or ions was thus treated by means of quantum mechanics very soon after its discovery.

Oppenheimer (1928) developed a method for computing the probabilities for transitions between states of the same energy, represented by almost orthogonal eigenfunctions, and applied the resulting formula to treat the ionization of hydrogen atoms in a homogeneous electric field. Somewhat later Lanczos (1930a, 1930b) treated the Stark effect for a hydrogen atom in a strong electric field by deriving an approximate asymptotic solution for the one of the previously mentioned ordinary differential equations that has a continuous energy spectrum. He pointed out that the Stark levels are not sharp but have a finite width which he discussed. Lanczos (1930c) also improved the method of asymptotic treatment of the Stark effect for a hydrogen atom, with the magnetic quantum number m equal to zero, in a strong electric field. The asymptotic method he used is closely related to the first order of the Carlini (JWKB) approximation along with Jeffreys' (1925) connection formulas for that approximation, the one-directional validity of which is, however, not discussed. For the positions of the Stark levels Lanczos arrived at a quantization condition of the Bohr–Sommerfeld type, which he expressed in terms of complete elliptic integrals of the first and second kind. He also discussed the breakdown of perturbation treatments for strong electric fields. On the basis partly of the time-independent and partly of the time-dependent Schrödinger equation, Lanczos (1931) discussed, although in a not quite clear way, the weakening of the intensities of the spectral lines and the ionization of atomic hydrogen in strong electric fields. For the disintegration constant he obtained an expression in terms of complete elliptic integrals of the first and second kind.

Publications with relevance to this book

We shall consider mainly publications in which asymptotic methods are used, but we also mention numerical methods, since we use numerical results for comparison with our phase-integral results. For a general review of the field we refer to Bethe and Salpeter (1957), Ryde (1976), Bayfield (1979), Koch (1981), Gallas, Leuchs, Walther and Figger (1985), Lisitsa (1987) and Gallagher (1988, 1994).

Rice and Good (1962) calculated the positions of the Stark levels of atomic hydrogen in a homogeneous electric field by using the Carlini (JWKB) approximation combined with comparison equation technique for the treatment of the time-independent Schrödinger equation separated in parabolic coordinates. They considered in particular the case when the energy lies close to the top of the barrier. For the positions of the energy levels the authors obtained quantization conditions expressed in terms of complete elliptic integrals of the first and second kind. Furthermore, they improved Lanczos' (1930b, 1930c, 1931) estimate of the dependence of the lifetime on field ionization and calculated also the half-width of the Stark levels. Thus they obtained formulas for the lifetime and the half-width of the Stark levels in terms of complete elliptic integrals of the first and second kind. Due to a need for explicit values of the ionization probabilities up to very high energy levels, several electric field ionization probabilities for a hydrogen atom in an electric field were calculated by Bailey, Hiskes and Riviere (1965) by the methods of Lanczos (1931) and Rice and Good (1962). The results were presented graphically and in a table. Guschina and Nikulin (1975) calculated the resonance energy and the decay probability for a particular quasistationary state of a hydrogen atom in a homogeneous electric field by numerical integration of the two coupled differential equations obtained by separation of the time-independent Schrödinger equation in parabolic coordinates. The values obtained for the resonance energy agree perfectly with values obtained by Rayleigh–Schrödinger perturbation theory up to the fourth power in the electric field strength and rather well with values obtained by Bailey, Hiskes and Riviere (1965). The values of the decay probability agree rather well with those obtained by Bailey, Hiskes and Riviere (1965). To solve the two differential equations, obtained by separation in parabolic coordinates of the Schrödinger equation for a hydrogenic atom in a homogeneous electric field, Bekenstein and Krieger (1969) used the Carlini (JWKB) approximation and derived quantization conditions in the fifth order of that approximation. From these quantization conditions the authors obtained for the positions of the Stark energy levels a series up to the fourth power of the electric field strength. This series agrees, for those states for which comparison could be made, with the corresponding series obtained by perturbation theory. The general conclusion of Bekenstein and Krieger seems to be that the use of the Carlini (JWKB) approximation is superior to the use of perturbation theory for all Stark levels of a hydrogenic atom. Alliluev and Malkin (1974) derived the perturbation series for the Stark effect of atomic hydrogen up to the fourth power of the electric field strength. They find that their result is in complete agreement with the results of previous authors up to the third-order correction. Although they find a disagreement in their fourth-order correction with the result obtained by Bekenstein and Krieger (1969), they express the opinion that the correct Carlini (JWKB) approximation and perturbation theory lead to identical results in the case of weak electric fields. Furthermore, Alliluev and Malkin (1974) quote Basu's (1934) result for the fourth-order correction, which is published in a journal that is almost inaccessible, and point out that his fourth-order formula contains errors. Herrick (1976) confirms on page 3534 that Alliluev and Malkin (1974) corrected errors in both the Basu (1934) formula and in the WKB expansion of Bekenstein and Krieger (1969). Yamabe, Tachibana and Silverstone (1977) developed the theory of the ionization of a hydrogen atom in an electric field analytically and corrected Oppenheimer's (1928) formula for the ionization in a weak electric field. As a general conclusion of this and other results, the authors state that the field ionization of hydrogen is unsuspectedly insidious, having left a legacy of errors. Drukarev (1978) calculated in the quasiclassical approximation the energies and widths of energy levels of a hydrogen atom in a homogeneous electric field. Later Drukarev (1982) considered the Stark effect when the energy level lies at the top of the barrier. Gallas, Walther and Werner (1982a) used the first-order Carlini (JWKB) approximation to treat the Stark effect in a hydrogen atom for arbitrary values of the magnetic quantum number m. In the two coupled, ordinary differential equations, obtained after separation in parabolic coordinates, these authors erroneously replaced $m^2 - 1$ by m^2 and obtained differential equations that are not correct. Many other authors have also made this serious mistake, and therefore it is important to emphasize that the replacement of $m^2 - 1$ by m^2, or $l(l+1)$ by $(l+1/2)^2$ in a radial problem, is not to be made in the *differential equations* but only in the *first-order Carlini (JWKB) approximation*, and that this replacement in the higher-order corrections does not give a correct result. For the positions of the Stark levels well below the top of the barrier the authors obtained quantization conditions expressed in terms of complete elliptic integrals of the first, second and third kind, which they extended in an unsatisfactory way to energy levels above the top of the barrier. Somewhat later Gallas, Walther and Werner (1982b) used the first-order Carlini (JWKB) approximation and handled the three-turning-point problem also when the energy may lie close to the top of the barrier, but they made the same mistake as in their previous paper (1982a). For the ionization rate of a hydrogenic atom or ion in an electric field they obtained a simple formula, expressed in terms of complete elliptic integrals of the first, second and third kind, which they found to be in excellent agreement with results obtained from numerically exact calculations, and which for energies well below the top of the barrier agrees with the formula obtained by Rice and Good (1962). On the basis of, on the one hand the first-order Carlini (JWKB) approximation combined with comparision equation results, and on the other hand a purely numerical method, Farrelly and Reinhardt (1983) performed calculations of complex energy eigenvalues for a hydrogen atom in a homogeneous electric field. They demonstrated the efficiency and remarkable accuracy of the Carlini (JWKB) approximation already in the first-order approximation and pointed out that previous discrepancies between results obtained by the use of that approximation and accurate numerical results, which had usually been attributed to the break-down of the approximation, are rather due to a failure to use the approximation in a correct and uniform way. The authors concluded that an appropriate approach based on the approximation in question is an efficient and highly accurate method for the calculation of complex energy eigenvalues for the Stark problem. Korsch and Möhlenkamp (1983) performed, independently of Farrelly and Reinhardt (1983), a similar investigation. By means of comparison equation technique Kolosov (1983) determined the energy and the ionization probability of a hydrogen atom in a homogeneous electric field, when the energies of the differential equation describing tunneling through the potential barrier lie near the top of the barrier. Formulas for the energy and the ionization probability in some previous papers are characterized as either erroneous or too complicated.

Though the exact solution of the Stark effect problem for hydrogenic atoms or ions can in principle be obtained by numerical integration of the two ordinary differential equations, resulting from the separation of the three-dimensional Schrödinger equation in parabolic coordinates, exact calculations encounter computational difficulties and have hence been rather few [see, however, Alexander (1969) and Hirschfelder and Curtiss (1971)] until the extensive calculations of positions and widths of Stark levels by Damburg and Kolosov (1976a, 1976b, 1977, 1978a, 1978b, 1979, 1980, 1981, 1982) and by Kolosov (1983, 1987) began to appear. A numerical method for calculating normalized wave functions and absolute values for the densitiy of oscillator strengths in the photoabsorption spectrum of hydrogenic atoms or ions in the presence of a homogeneous electric field has been presented by Luc-Koenig and Bachelier (1980a,b).

Treatment in this book

In the present book we treat the Stark effect for a hydrogenic atom or ion in a homogeneous electric field with the use of the phase-integral approximation generated from a conveniently chosen base function; see for a detailed presentation of that approximation Chapter 1 in Fröman and Fröman (1996) and for a summary Section 4.1 in the present book. We shall give a non-relativistic treatment of the problem in which electron spin, fine structure and hyperfine structure are not taken into account. Furthermore, we assume that the time for electric field ionization due to the Stark effect is much smaller than the time for emission of a photon from the state in question. Some previous authors have combined the Carlini (JWKB) approximation and comparison equation technique; see for instance Rice and Good (1962), Bailey, Hiskes and Riviere (1965) and Harmin (1981). We do not proceed in a corresponding way, since comparison equation technique has already been used to obtain the general, analytic, arbitrary-order phase-integral formulas on which we base our treatment of the Stark effect; see Fröman and Fröman (1996). In particular we use an arbitrary-order phase-integral formula for barrier transmission (Fröman and Fröman 2002), which allows the energy to lie close to and even above the top of the barrier. Our treatment is thus in several respects more satisfactory and more straightforward than previous asymptotic treatments. Finally we arrive at phase-integral formulas, expressed in terms of complete elliptic integrals of the first, second and third kind, for Stark level profiles, positions and half-widths.

We share the opinion expressed by Farrelly and Reinhardt (1983) that discrepancies between Stark effect results obtained by the use of the Carlini (JWKB) approximation and by accurate numerical calculations cannot be attributed to the break-down of the approximation, but are due to a failure to use the approximation in a correct way. An appropriate approach based on the phase-integral approximation of arbitrary order generated from an appropriately chosen base function is a still more efficient and often highly accurate method for the treatment of several problems, not only in quantum mechanics, but in various fields of theoretical physics. With great success it has for instance been used in the study of black-hole normal modes (Fröman, Fröman, Andersson and Hökback 1992; Andersson, Araújo and Schutz 1993) and in the study of cosmological perturbations during inflation (Rojas and Villalba 2007). See also Athavan *et al.* (2001a–c), which concerns the two-center Coulomb problem, and where the reason for the possibility of obtaining their accurate results is the presence of the unspecified base function from which the phase-integral approximation is generated. Such accurate results cannot be obtained by means of the Carlini (JWKB) approximation, since there is no unspecified base function in that approximation.

We shall now illustrate the accuracy of the energy values obtained by means of our phase-integral formulas. For 198 different Stark states of a hydrogen atom, with either different quantum numbers or the same quantum numbers but different electric field strengths, we present in the tables in Chapter 8 values of the energy and the half-width that have been calculated by means of the phase-integral approximation generated from an appropriate base function as well as by other methods. We emphasize that all results there have been obtained by neglecting fine structure corrections. Compared to the best energy values obtained by other methods, the optimum phase-integral energy values are for these states judged to be at least as accurate in more than half of the cases; see Fig. 1.1. The phase-integral formulas can sometimes give results of surprisingly great accuracy. Compared to the numerically obtained results, the phase-integral results in Chapter 8 can contain up to seven more digits for the energy eigenvalues. A more detailed presentation of the accuracy of the phase-integral energies versus the accuracy of the energies obtained by other methods is given in Fig. 1.1.

For large field strengths (thin barriers) the phase-integral method gives usually better results than for small field strengths (thick barriers). The phase-integral method is therefore an important complement to the numerical methods, which are in general less accurate for large field strengths than for small field strengths. For very thick barriers the numerical methods do not give good values of the half-widths, and for extremely thick barriers they may sometimes only give upper limits for the half-widths, while the phase-integral method

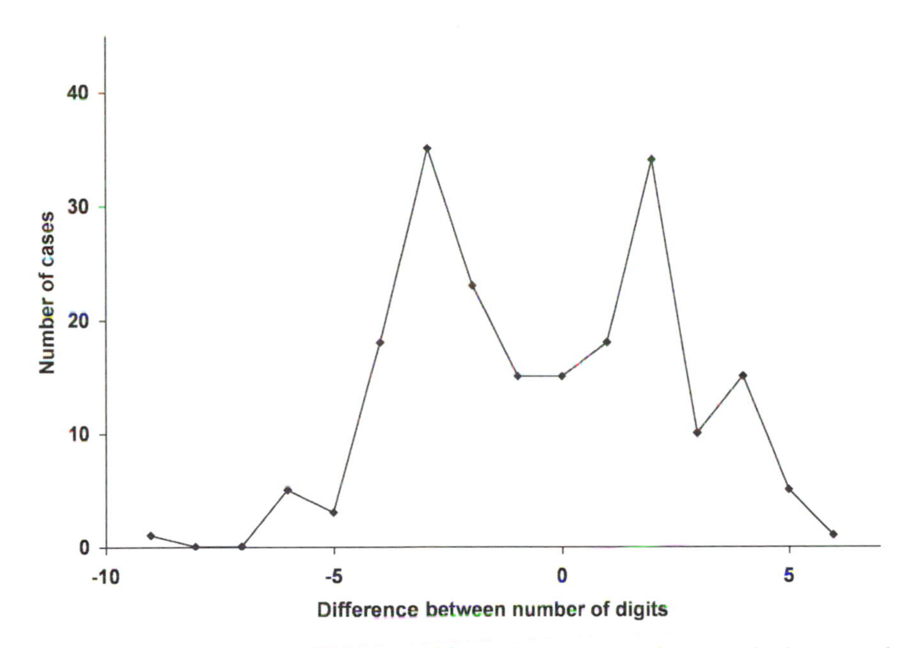

Fig. 1.1. For the 198 cases in the tables in Chapter 8 we have on the horizontal axis in this figure plotted the difference between the number of correct digits in the optimum phase-integral energy value, obtained with $\tilde{\phi}$ included, and the number of correct digits in the best energy value obtained by other methods, while on the vertical axis we have plotted the number of cases corresponding to different values of this difference. Recalling a property of asymptotic series, we have considered the error of the optimum phase-integral energy values E to be of the order of the smallest difference between the E-values for two consecutive optimum orders of the phase-integral approximation. When the phase-integral values of E are judged to be less accurate than the other values of E, we have in general considered all digits in the best of the other values to be correct. Since there is certainly an unknown number of exceptions from this assumption, the part of the figure that lies to the left of the origin underestimates the accuracy of the phase-integral method. According to this figure there are 98 cases in which the energy values obtained by the phase-integral method are at least as accurate as those obtained by other methods, but because of what has just been said, we believe that the energy values are obtained at least as accurately by the phase-integral method as by other methods in more than half of the cases.

gives rather accurate values; see Tables 8.10c, 8.10f and 8.10l in Chapter 8. For large values of the quantum number n_2 the phase-integral method gives often more accurate results than for small values of n_2.

For some time there was a puzzling discrepancy between a phase-integral result obtained by us and a corresponding result obtained numerically by Damburg and Kolosov, and quoted by Silverstone and Koch (1979), for the position of the Stark level of a hydrogen atom with the quantum numbers $m = 0$, $n_1 = 0, n_2 = 29$ and $n = |m| + 1 + n_1 + n_2 = 30$ in a homogeneous electric field of the strength 800 V/cm. This is a state with the first-order barrier penetration phase-integral K (see Section 4.3) close to unity. It is thus situated not very far below the top of the barrier. Anders Hökback, who was research engineer at the Department of Theoretical Physics, University of Uppsala, found the explanation of this discrepancy, which turned out to be due to the seemingly innocent approximation of considering L, K and $\tilde{\phi}$, defined in (5.15a,b), (5.19a,b) and (4.46)–(4.48a,b), as constants over the width of the level, i.e., to use the quantization condition (5.33). When we avoided this approximation and used the more accurate quantization condition (5.44) and the formula (5.42) for Δ, the discrepancy disappeared completely. The state in question has been discussed in detail by Silverstone and Koch (1979), who reported results obtained by 24th-order Rayleigh–Schrödinger perturbation theory (RSPT) combined with a (12/12) Padé approximant (PA). For the same state Damburg and Kolosov improved their above-mentioned result by computing a more accurate numerical result (private communication from Damburg to Nanny Fröman in a letter dated 22 February 1985). The position of the Stark level in question is calculated by us by means of the accurate formula (5.44) with Δ given by (5.42), i.e., with the energy dependence of L, K and $\tilde{\phi}$ over the width of the level taken into account. Our results and the above-mentioned results obtained by Silverstone and Koch (1979) and by Damburg and Kolosov are presented in Table 1.1, which (except for the improved Damburg–Kolosov energy value obtained as private communication) is part of a table that was presented by Fröman and Fröman (1984) at a conference. It is seen that already the first-order phase-integral result is slightly more accurate than the result obtained by Silverstone and Koch (1979) by the use of 24th-order Rayleigh–Schrödinger perturbation theory combined with a (12/12) Padé approximant. Our third- and fifth-order energy values are at least as accurate as the value obtained

Table 1.1. The data in this table, which have been taken from Table 8.7 in Chapter 8, refer to a hydrogen atom with the quantum numbers $m = 0$, $n_1 = 0, n_2 = 29$ and $n = |m| + 1 + n_1 + n_2 = 30$ in an electric field of the strength 800 V/cm. The value called RSPT + PA has been obtained by Silverstone and Koch (1979), who remarked that the two underlined digits are uncertain. In the value called Numerical, which has been obtained by Damburg and Kolosov (private communication from Damburg to Nanny Fröman in a letter dated 22 February 1985) there is some doubt about the underlined digit. Like Damburg and Kolosov we have used the conversion factor 1 au = 5.142 260 3 V/cm.

Method of calculation	$-E \times 10^4$ au	$\Gamma \times 10^7$ au
1st-order phase-integral approximation	7.844 656	2.849
3rd-order	7.844 648 053	2.853 2
5th-order	7.844 648 046	2.853 2
Numerical (Damburg and Kolosov)	7.844 648 0$\underline{4}$	
RSPT + PA (Silverstone and Koch)	7.844 $\underline{68}$	

numerically by Damburg and Kolosov. The agreement between our results in the third and fifth orders of approximation indicates in fact that the last digit in the value obtained by Damburg and Kolosov may be wrong by one unit in the last digit. The last digit in the value obtained by Silverstone and Koch (1979) is wrong by three units. The number of reliable digits for E obtained by the phase-integral method is nine. We emphasize that the results in Table 1.1 have been obtained by disregarding the fine structure corrections, which may be of the order of 10^{-6} to 10^{-5}. Therefore the results in this table that are extremely accurate do not represent experimental reality; they are only intended to show the accuracy obtainable by different methods of calculation. There does not seem to exist an experimental value for the energy of a hydrogen atom with the quantum numbers $m = 0$, $n_1 = 0$, $n_2 = 29$ in an electric field with the strength 800 V/cm.

Brief account of the contents of this book

In Chapter 2 the time-dependent Schrödinger equation, describing the Stark effect of a hydrogenic atom or ion in a homogeneous electric field, is separated with respect to time dependence, center of mass motion, and internal motion. The time-independent Schrödinger equation for the internal motion is then separated in parabolic coordinates. The result is a system of two coupled differential equations, one with a discrete energy spectrum, and the other with a continuous energy spectrum. Enclosing the independent variable η of the differential equation with a continuous energy spectrum in the large but finite interval, $0 \leq \eta \leq \rho$, i.e., imposing on the wave function $g(\eta)$ the boundary conditions $g(0) = 0$ and $g(\rho) = 0$, we perform an analysis of the properties of the eigenfunctions of the two coupled differential equations. In Chapter 3 we consider the development in time of the wave function for the internal motion. The result is an exact formula for the probability amplitude of a decaying state. The eigenfunctions of the above-mentioned two coupled differential equations appear in this formula, which provides the basis for the further treatment of the hydrogenic Stark effect by means of the phase-integral approximation generated from an appropriate base function. With the use of this approximation, which is briefly described in Chapter 4, we obtain in Chapter 5 a more explicit expression for the development in time of the probability amplitude of a decaying state. This expression, which is obtained in the limit $\rho \to \infty$, contains an energy-dependent quantity $(\Omega'/\Omega'')^2$, which can be interpreted as the level profile. It yields a natural definition of the position and (when the spectral line is not too broad) of the half-width of the Stark level. In Chapter 6 it is described how one transforms the phase-integral formulas derived in Chapter 5 into formulas expressed in terms of complete elliptic integrals of the first, second and third kind. The formulas thus obtained are collected in Chapter 7. These formulas along with well-known properties of complete elliptic integrals, such as for instance series expansions, can be exploited for analytic studies of the Stark effect. Complete elliptic integrals can be evaluated very rapidly by means of standard computer programs, and with the use of the formulas in Chapter 7 a comprehensive numerical material concerning the Stark effect of atomic hydrogen has been obtained. It is presented in Chapter 8, where positions and half-widths for various levels are compared with corresponding results reported by other authors.

Chapter 2

Schrödinger Equation, its Separation and its Exact Eigenfunctions

When a hydrogenic atom or ion, in which the nucleus has the charge $Ze(e > 0)$, the mass μ_1 and the position $\vec{r}_1 = (x_1, y_1, z_1)$, and the electron has the charge $-e(e > 0)$, the mass μ_2 and the position $\vec{r}_2 = (x_2, y_2, z_2)$, is placed in a homogeneous electric field of the strength $\overline{F}(>0)$ and with the direction of the *positive* z-axis, the Hamiltonian of this system is

$$H = -\frac{\hbar^2}{2\mu_1}\Delta_{\vec{r}_1} - \frac{\hbar^2}{2\mu_2}\Delta_{\vec{r}_2} - \frac{Ze^2}{|\vec{r}_1 - \vec{r}_2|} - Ze\overline{F}z_1 + e\overline{F}z_2, \tag{2.1}$$

if we do not take into account relativistic effects, spin and the fine structure of the hydrogenic energy levels. The imposed electric field $\overline{F}$ is thus assumed to be so strong that the Stark splitting is large compared to the fine structure splitting. Introducing the position vector $\vec{r}_0 = (x_0, y_0, z_0)$ for the center of mass, i.e.,

$$\vec{r}_0 = \frac{\mu_1\vec{r}_1 + \mu_2\vec{r}_2}{\mu_1 + \mu_2}, \tag{2.2}$$

and the relative position vector $\vec{r} = (x, y, z)$ of the electron, i.e.,

$$\vec{r} = \vec{r}_2 - \vec{r}_1, \tag{2.3a}$$

$$r = |\vec{r}_2 - \vec{r}_1|, \tag{2.3b}$$

one can write (2.1) as

$$\begin{aligned} H = &-\left[\frac{\hbar^2}{2(\mu_1+\mu_2)}\Delta_{\vec{r}_0} + (Z-1)e\overline{F}z_0\right] \\ &-\left[\frac{\hbar^2}{2\mu}\Delta_{\vec{r}} + \frac{Ze^2}{r} - \frac{\mu_1 + Z\mu_2}{\mu_1+\mu_2}e\overline{F}z\right], \end{aligned} \tag{2.4}$$

where

$$\mu = \frac{\mu_1\mu_2}{\mu_1+\mu_2} \tag{2.5}$$

is the reduced mass. The Schrödinger equation for the system is

$$H\Psi = -\frac{\hbar}{i}\frac{\partial\Psi}{\partial t}. \tag{2.6}$$

Putting

$$\Psi = \chi_0(\vec{r}_0)\chi(\vec{r})T(t) \tag{2.7}$$

and recalling (2.4), we can write (2.6) as

$$\begin{aligned} &-\frac{1}{\chi_0(\vec{r}_0)}\left[\frac{\hbar^2}{2(\mu_1+\mu_2)}\Delta_{\vec{r}_0} + (Z-1)e\overline{F}z_0\right]\chi_0(\vec{r}_0) \\ &-\frac{1}{\chi(\vec{r})}\left(\frac{\hbar^2}{2\mu}\Delta_{\vec{r}} + \frac{Ze^2}{r} - \frac{\mu_1+Z\mu_2}{\mu_1+\mu_2}e\overline{F}z\right)\chi(\vec{r}) = -\frac{\hbar}{iT(t)}\frac{dT(t)}{dt}. \end{aligned} \tag{2.8}$$

Since each term in (2.8) must be equal to a constant, we put

$$-\frac{1}{\chi_0(\vec{r}_0)}\left[\frac{\hbar^2}{2(\mu_1+\mu_2)}\Delta_{\vec{r}_0} + (Z-1)e\overline{F}z_0\right]\chi_0(\vec{r}_0) = E_0, \tag{2.9a}$$

$$-\frac{1}{\chi(\vec{r})}\left(\frac{\hbar^2}{2\mu}\Delta_{\vec{r}} + \frac{Ze^2}{r} - \frac{\mu_1+Z\mu_2}{\mu_1+\mu_2}e\overline{F}z\right)\chi(\vec{r}) = E, \tag{2.9b}$$

$$-\frac{\hbar}{iT(t)}\frac{dT(t)}{dt} = E_0 + E. \tag{2.9c}$$

The physically relevant solution of (2.9a), which represents the motion of the center of mass, is when Z=1

$$\begin{aligned} \chi_0(\vec{r}_0) &= \text{const} \times \exp(ik_x x_0 + ik_y y_0 + ik_z z_0) \\ &= \text{const} \times \exp(i\vec{k}\cdot\vec{r}_0), \quad Z = 1, \end{aligned} \tag{2.10a}$$

where

$$k_x^2 + k_y^2 + k_z^2 = \frac{2(\mu_1 + \mu_2)E_0}{\hbar^2}, \tag{2.11a}$$

and when $Z \neq 1$

$$\chi_0(\vec{r}_0) = \text{const} \times \exp(ik_x x_0 + ik_y y_0) Ai\left(-\kappa z_0 - \frac{k_z^2}{\kappa^2}\right), \quad Z \neq 1, \tag{2.10b}$$

where

$$\kappa = \left[\frac{2(\mu_1 + \mu_2)(Z-1)eF}{\hbar^2}\right]^{1/3}. \tag{2.11b}$$

The solution of (2.9c) is

$$T(t) = \text{const} \times \exp\left[-\frac{i(E_0 + E)t}{\hbar}\right]. \tag{2.12}$$

Except for a constant normalization factor the solution (2.7) can thus be written as

$$\Psi = \psi_0(\vec{r}_0, t)\psi(\vec{r}, t), \tag{2.13}$$

where $\psi_0(\vec{r}_0, t)$ is given by either of the formulas

$$\begin{aligned} \psi_0(\vec{r}_0, t) &= \exp(ik_x x_0 + ik_y y_0 + ik_z z_0) \exp\left(-\frac{iE_0 t}{\hbar}\right) \\ &= \exp\left(i\vec{k} \cdot \vec{r}_0 - \frac{iE_0 t}{\hbar}\right), \quad Z = 1, \end{aligned} \tag{2.14a}$$

$$\begin{aligned} \psi_0(\vec{r}_0, t) &= \exp(ik_x x_0 + ik_y y_0) \\ &\quad \times Ai\left(-\kappa z_0 - \frac{k_z^2}{\kappa^2}\right) \exp\left(-\frac{iE_0 t}{\hbar}\right), \quad Z \neq 1, \end{aligned} \tag{2.14b}$$

with E_0 and κ obtained from (2.11a) and (2.11b), and $\psi(\vec{r}, t)$ is given by

$$\psi(\vec{r}, t) = \chi(\vec{r}) \exp\left(-\frac{iEt}{\hbar}\right), \tag{2.15}$$

$\chi(\vec{r})$ being a solution of the differential equation (2.9b), i.e., the time-independent Schrödinger equation for the internal motion:

$$\left(-\frac{\hbar^2}{2\mu}\Delta_{\vec{r}} - \frac{Ze^2}{r} + eFz\right)\chi = E\chi, \tag{2.16}$$

where

$$F = \frac{\mu_1 + Z\mu_2}{\mu_1 + \mu_2}\overline{F} \tag{2.17}$$

is the effective electric field strength. We note that the effective electric field F is equal to the imposed electric field $\overline{F}$ for a hydrogen atom ($Z = 1$) but slightly different from $\overline{F}$ for a hydrogenic ion ($Z \neq 1$).

2.1 Separation of the time-independent Schrödinger equation for the internal motion

In a well-known way we introduce the parabolic coordinates $\xi(\geq 0), \eta(\geq 0)$ and φ by writing

$$x = (\xi\eta)^{1/2}\cos\varphi, \tag{2.18a}$$

$$y = (\xi\eta)^{1/2}\sin\varphi, \tag{2.18b}$$

$$z = \frac{1}{2}(\xi - \eta). \tag{2.18c}$$

Hence

$$r = (x^2 + y^2 + z^2)^{1/2} = \frac{1}{2}(\xi + \eta). \tag{2.19}$$

We obtain from (2.18c) and (2.19)

$$\xi = r + z, \quad 0 \leq \xi < \infty, \tag{2.20a}$$

$$\eta = r - z, \quad 0 \leq \eta < \infty, \tag{2.20b}$$

and from (2.18a) and (2.18b)

$$\varphi = \arctan\left(\frac{y}{x}\right). \tag{2.20c}$$

We also note that

$$\frac{\partial(x, y, z)}{\partial(\xi, \eta, \varphi)} = \frac{1}{4}(\xi + \eta) \tag{2.21}$$

and hence

$$dx\, dy\, dz = \frac{1}{4}(\xi + \eta) d\xi\, d\eta\, d\varphi. \tag{2.22}$$

To solve the time-independent Schrödinger equation (2.16) we put

$$\chi = \Omega \frac{f(\xi)}{\xi^{1/2}} \frac{g(\eta)}{\eta^{1/2}} \Phi(\varphi), \tag{2.23}$$

where $\Omega(\neq 0)$ is a normalization factor, which we for the sake of simplicity assume to be real, and note that

$$\Delta_{\vec{r}} = \frac{4}{\xi+\eta}\left(\xi^{1/2}\frac{\partial^2}{\partial\xi^2}\xi^{1/2} + \frac{1}{4\xi} + \eta^{1/2}\frac{\partial^2}{\partial\eta^2}\eta^{1/2} + \frac{1}{4\eta}\right) + \frac{1}{\xi\eta}\frac{\partial^2}{\partial\varphi^2}. \tag{2.24}$$

Recalling (2.18c), (2.19), (2.23) and (2.24), we can write (2.16) as

$$\frac{4\xi\eta}{\xi+\eta}\left[\frac{\xi}{f(\xi)}\frac{d^2 f(\xi)}{d\xi^2} + \frac{1}{4\xi} + \frac{\eta}{g(\eta)}\frac{d^2 g(\eta)}{d\eta^2} + \frac{1}{4\eta}\right] + \xi\eta\left[\frac{4\mu Ze^2}{\hbar^2(\xi+\eta)} - \frac{\mu eF(\xi-\eta)}{\hbar^2} + \frac{2\mu E}{\hbar^2}\right] + \frac{1}{\Phi(\varphi)}\frac{d^2\Phi(\varphi)}{d\varphi^2} = 0 \tag{2.25}$$

and hence

$$\frac{1}{\Phi(\varphi)}\frac{d^2\Phi(\varphi)}{d\varphi^2} = -m^2, \tag{2.26a}$$

$$\frac{4\xi\eta}{\xi+\eta}\left[\frac{\xi}{f(\xi)}\frac{d^2 f(\xi)}{d\xi^2} + \frac{1}{4\xi} + \frac{\eta}{g(\eta)}\frac{d^2 g(\eta)}{d\eta^2} + \frac{1}{4\eta}\right] + \xi\eta\left[\frac{4\mu Ze^2}{\hbar^2(\xi+\eta)} - \frac{\mu eF(\xi-\eta)}{\hbar^2} + \frac{2\mu E}{\hbar^2}\right] = m^2, \tag{2.26b}$$

where m^2 is a separation constant. We can write (2.26b) as

$$\frac{\xi}{f(\xi)}\frac{d^2 f(\xi)}{d\xi^2} - \frac{\mu eF\xi^2}{4\hbar^2} + \frac{\mu E\xi}{2\hbar^2} + \frac{1-m^2}{4\xi} + \frac{\eta}{g(\eta)}\frac{d^2 g(\eta)}{d\eta^2} + \frac{\mu eF\eta^2}{4\hbar^2} + \frac{\mu E\eta}{2\hbar^2} + \frac{1-m^2}{4\eta} = -\frac{\mu Ze^2}{\hbar^2}, \tag{2.27}$$

and from this equation it follows that

$$\frac{\xi}{f(\xi)}\frac{d^2 f(\xi)}{d\xi^2} - \frac{\mu eF\xi^2}{4\hbar^2} + \frac{\mu E\xi}{2\hbar^2} + \frac{1-m^2}{4\xi} = -Z_1, \tag{2.28a}$$

$$\frac{\eta}{g(\eta)}\frac{d^2 g(\eta)}{d\eta^2} + \frac{\mu eF\eta^2}{4\hbar^2} + \frac{\mu E\eta}{2\hbar^2} + \frac{1-m^2}{4\eta} = -Z_2, \tag{2.28b}$$

where Z_1 and Z_2 are separation constants subjected to the condition

$$Z_1 + Z_2 = \frac{\mu Ze^2}{\hbar^2}. \tag{2.29}$$

For hydrogen atoms ($Z = 1$) and with units such that $\mu = e = \hbar = 1$ one sees from the tables in Chapter 8 that $0 < Z_1 < 1$, and hence it follows from (2.29) that $0 < Z_2 < 1$. With the further restriction that $F = 0$ Yamabe, Tachibana and Silverstone (1977) give in their Eqs. (13)–(15) analytical expressions for Z_1 and Z_2 in terms of η_1, η_2 and $|m|$. The equations (2.26a), (2.28a) and (2.28b) can be written as

$$\frac{d^2\Phi}{d\varphi^2} + m^2\Phi = 0, \quad 0 \leq \varphi \leq 2\pi, \tag{2.30}$$

$$\frac{d^2 f}{d\xi^2} + \tilde{R}(\xi) f = 0, \quad 0 \leq \xi < \infty, \tag{2.31a}$$

$$\tilde{R}(\xi) = \frac{\mu E}{2\hbar^2} + \frac{Z_1}{\xi} + \frac{1 - m^2}{4\xi^2} - \frac{\mu e F \xi}{4\hbar^2}, \tag{2.31b}$$

$$\frac{d^2 g}{d\eta^2} + R(\eta) g = 0, \quad 0 \leq \eta < \infty, \tag{2.32a}$$

$$R(\eta) = \frac{\mu E}{2\hbar^2} + \frac{Z_2}{\eta} + \frac{1 - m^2}{4\eta^2} + \frac{\mu e F \eta}{4\hbar^2}. \tag{2.32b}$$

The general solution of (2.30) is an arbitrary linear combination of the functions $\exp(\pm im\varphi)$. Since the wave function χ must be single-valued when φ changes by 2π, the only possible values of m are the integers $m = 0, \pm 1, \pm 2, \ldots$.

The behavior of the solutions $f(\xi)$ and $g(\eta)$ of the differential equations (2.31a,b) and (2.32a,b), respectively, for small values of ξ and η, respectively, can be found in a well-known way by means of the indicial equation.

For $m \neq 0$ the result is that there are solutions $f(\xi)$ and $g(\eta)$ that for small values of ξ and η are approximately proportional to $\xi^{(1+|m|)^2}$ and $\eta^{(1+|m|)^2}$, respectively. Hence $f(\xi)/\xi^{1/2}$ and $g(\eta)/\eta^{1/2}$ are approximately proportional to $\xi^{\pm|m|/2}$ and $\eta^{\pm|m|/2}$, respectively. Since we require χ, which is given by (2.23), to be finite everywhere, we do not accept the minus signs in these expressions. Therefore the physically acceptable solutions $f(\xi)$ and $g(\eta)$ are for small values of ξ and η equal to $\xi^{(1+|m|)^2}$ and $\eta^{(1+|m|)^2}$, respectively, times a power series in ξ and η, respectively, with the constant term different from zero.

For $m = 0$ the indicial equation yields one function $f(\xi)/\xi^{1/2}$ that is approximately equal to a constant for small values of ξ, and one finds that there is another function $f(\xi)/\xi^{1/2}$ that is approximately proportional to $\ln \xi$ for small values of ξ. Similarly one obtains two functions $g(\eta)/\eta^{1/2}$ with corresponding behavior for small values of η. Of these functions we accept only those that are approximately equal to a constant (different from zero) for small values of ξ and η, respectively, since otherwise χ would not be finite everywhere.

For any allowed value of m, i.e., for $m = 0, \pm 1, \pm 2, \ldots$, one thus obtains the physically acceptable wave function χ, which is finite everywhere, from the particular solution $f(\xi)$ that is equal to $\xi^{(1+|m|)^2}$ times a power series in ξ with the constant term different from zero and from the particular solution $g(\eta)$ that is equal to $\eta^{(1+|m|)^2}$ times a power series in η with the constant term different from zero. These particular solutions $f(\xi)$ and $g(\eta)$, which are single-valued and uniquely determined except for arbitrary constant factors, obviously tend to zero as $\xi \to 0$ and $\eta \to 0$, respectively.

The differential equation (2.31a,b) for $f(\xi)$ has the form of a radial Schrödinger equation for a particle in a potential well. For given values of m and Z_1 there is therefore a series of discrete values of E (characterized by the quantum number $n_1 = 0, 1, 2, \ldots$) for which this differential equation has acceptable solutions. On the other hand, if $F \neq 0$ (which we shall assume from now on) the differential equation (2.32a,b) for $g(\eta)$ is the same as that for a particle that can penetrate a potential barrier, and hence this differential equation has, for given values of m and Z_2, physically acceptable solutions for all possible values of E. It is convenient to confine the system in a region such that $0 \leq \xi < \infty$ and $0 \leq \eta \leq \rho$, where ρ is a large positive number, which we shall finally let it tend to infinity. From (2.20a,b) it follows that this confinement in the η-space corresponds to the confinement $z \geq (x^2 + y^2)/(2\rho) - \rho/2$ in the xyz-space. When ρ is finite, we impose on $g(\eta)$ the condition that $g(\rho) = 0$. For given values of F, m and $Z_2[= \mu Z e^2/\hbar^2 - Z_1$ according to (2.29)] we then get a discrete series of very closely spaced E-values (characterized by the quantum number $s = 0, 1, 2, \ldots$). Thus, for a given effective field strength F and given quantum numbers m, n_1, and s one obtains from (2.31a,b) E as a function of Z_1, and one obtains from (2.32a,b) E as a function of $Z_2(= \mu Z e^2/\hbar^2 - Z_1)$. The requirement that these two energy eigenvalues coincide, together with the relation (2.29) between Z_1 and Z_2 then determines the values of E, Z_1 and Z_2. The relation (5.25) in Chapter 5 shows that for given values of F, m and n_1 the discrete values of E, like those of Z_1 and Z_2, are closely spaced. Since E increases as the quantum number s that determines the discrete values of E increases, we can denote the eigenvalues of E by $E_{m,n_1,s}$, the corresponding values of Z_1 and Z_2 by $Z_1(m, n_1, E_{m.n_1.s})$ and $Z_2(m, n_1.E_{m,n_1,s})$, and the eigenfunctions of (2.31a,b) and (2.32a,b) along with (2.29) by $f(m, n_1, E_{m,n_1,s}; \xi)$ and $g(m, n_1.E_{m,n_1,s}; \eta)$. These eigenfunctions are chosen to be real. The corresponding eigenfunction (2.23) is denoted by $\chi(m, n_1, E_{m,n_1,s}; x, y, z)$. Since the normalized eigenfunction $\Phi(\varphi)$ is equal to $\exp(im\varphi)/(2\pi)^{1/2}$, we can write (2.23) as

$$\chi(m, n_1, E_{m,n_1,s}; x, y, z) = \Omega(m, n_1, E_{m,n_1,s})\overline{\chi}(m, n_1, E_{m.n_1.s}; x, y, z) \tag{2.33}$$

with

$$\bar{\chi}(m, n_1, E_{m,n_1,s}; x, y, z) = \frac{f(m, n_1, E_{m,n_1,s}; \xi)}{\xi^{1/2}} \frac{g(m, n_1, E_{m,n_1,s}; \eta)}{\eta^{1/2}} \frac{\exp(im\varphi)}{(2\pi)^{1/2}}, \tag{2.34}$$

where m is an integer (positive, negative or zero), n_1 and s are non-negative integers, and $E_{m,n_1,s}$ is a discrete set of energy eigenvalues, which depend on m, n_1 and s as well as on the large quantity ρ and are closely spaced with respect to the quantum number s.

2.2 Properties of the eigenfunctions of the time-independent Schrödinger equation for the internal motion

Considering two states with the quantum numbers m, n_1, s and m', n_1', s', we first note that since the functions $\exp(im\varphi)$ form an orthogonal set on the interval $0 \leq \varphi \leq$, it follows from (2.33), (2.34) and (2.22) that

$$\iiint \chi^*(m, n_1, E_{m,n_1,s}; x, y, z) \times \chi(m', n_1', E_{m',n_1',s'}; x, y, z) dx\, dy\, dz = 0 \quad \text{if } m \neq m'. \tag{2.35}$$

Using (2.22), (2.33) and (2.34), we obtain, since the normalization factor in (2.33) and the functions $f(m, n_1, E_{m,n_1,s}; \xi)$ and $g(m, n_1, E_{m,n_1,s}; \eta)$ are assumed to be real,

$$\begin{aligned}
&\int\int\int \chi^*(m, n_1, E_{m,n_1,s}; x, y, z)\chi(m, n_1', E_{m,n_1',s'}; x, y, z)dx\, dy\, dz \\
&\quad = \Omega(m, n_1, E_{m,n_1,s})\Omega(m, n_1', E_{m,n_1',s'}) \\
&\quad\quad \times \left[\frac{1}{4}\int_0^\infty f(m, n_1, E_{m,n_1,s}; \xi)f(m, n_1', E_{m,n_1',s'}; \xi)d\xi \right. \\
&\quad\quad \times \int_0^\rho g(m, n_1, E_{m,n_1,s}; \eta)g(m, n_1', E_{m,n_1',s'}; \eta)\frac{d\eta}{\eta} \\
&\quad\quad + \frac{1}{4}\int_0^\infty f(m, n_1, E_{m,n_1,s}; \xi)f(m, n_1', E_{m,n_1',s'}; \xi)\frac{d\xi}{\eta} \\
&\quad\quad \left. \times \int_0^\rho g(m, n_1, E_{m,n_1,s}; \eta)g(m, n_1', E_{m,n_1',s'}; \eta)d\eta \right].
\end{aligned} \tag{2.36}$$

We write the differential equation (2.31a,b) as

$$\begin{aligned}
&\frac{d^2 f(m, n_1, E_{m,n_1,s}; \xi)}{d\xi^2} \\
&\quad + \left[\frac{\mu E_{m,n_1,s}}{2\hbar^2} + \frac{Z_1(m, n_1, E_m, n_1, s)}{\xi} + \frac{1-m^2}{4\xi^2} - \frac{\mu e F \xi}{4\hbar^2}\right] \\
&\quad \times f(m, n_1, E_{m,n_1,s}; \xi) = 0,
\end{aligned} \tag{2.37a}$$

when the quantum numbers are m, n_1, s, and as

$$\begin{aligned}
&\frac{d^2 f(m, n_1', E_{m,n_1',s'}; \xi)}{d\xi^2} \\
&\quad + \left[\frac{\mu E_{m,n_1',s'}}{2\hbar^2} + \frac{Z_1(m, n_1', E_m, n_1', s)}{\xi} + \frac{1-m^2}{4\xi^2} - \frac{\mu e F \xi}{4\hbar^2}\right] \\
&\quad \times f(m, n_1', E_{m,n_1',s'}; \xi) = 0,
\end{aligned} \tag{2.37b}$$

when the quantum numbers are m, n_1', s. Multiplying (2.37a) by $f(m, n_1', E_{m,n_1',s'}; \xi)$ and (2.37b) by $f(m, n_1, E_{m,n_1,s}; \xi)$, and subtracting from each other the two equations thus obtained, we get

$$\left(\frac{\mu(E_{m,n_1,s} - E_{m,n_1',s'})}{2\hbar^2} + \frac{Z_1(m,n_1,E_{m,n_1,s}) - Z_1(m,n_1',E_{m,n_1',s'})}{\xi}\right)$$
$$\times f(m,n_1,E_{m,n_1,s};\xi) f(m,n_1',E_{m,n_1',s'};\xi)$$
$$= \frac{d}{d\xi}\left[f(m,n_1,E_{m,n_1,s};\xi)\frac{df(m,n_1',E_{m,n_1',s'};\xi)}{d\xi}\right.$$
$$\left. - f(m,n_1',E_{m,n_1',s'};\xi)\frac{df(m,n_1,E_{m,n_1,s};\xi)}{d\xi}\right]. \tag{2.38}$$

Integrating (2.38) from $\xi = 0$ to $\xi = \infty$, and recalling the boundary conditions imposed at $\xi = 0$ and $\xi = \infty$, we get

$$\frac{\mu}{2\hbar^2}(E_{m,n_1,s} - E_{m,n_1',s'})\int_0^\infty f(m,n_1,E_{m,n_1,s};\xi) f(m,n_1',E_{m,n_1',s'};\xi)d\xi$$
$$+ \left[Z_1(m,n_1,E_{m,n_1,s}) - Z_1(m,n_1',E_{m,n_1',s'})\right]\int_0^\infty f(m,n_1,E_{m,n_1,s};\xi)$$
$$\times f(m,n_1',E_{m,n_1',s'};\xi)\frac{d\xi}{\xi} = 0, \tag{2.39}$$

i.e.,

$$\int_0^\infty f(m,n_1,E_{m,n_1,s};\xi) f(m,n_1',E_{m,n_1',s'};\xi)d\xi$$
$$= -\frac{2\hbar^2[Z_1(m,n_1,E_{m,n_1,s}) - Z_1(m,n_1',E_{m,n_1',s'})]}{\mu(E_{m,n_1,s} - E_{m,n_1',s'})}$$
$$\times \int_0^\infty f(m,n_1,E_{m,n_1,s};\xi) f(m,n_1',E_{m,n_1',s'};\xi)\frac{d\xi}{\xi}$$
$$\text{if} \quad E_{m,n_1,s} \neq E_{m,n_1',s'}. \tag{2.40}$$

From the differential equation (2.32a,b) we similarly obtain

$$\int_0^\rho g(m,n_1,E_{m,n_1,s};\eta) g(m,n_1',E_{m,n_1',s'};\eta)d\eta$$
$$= -\frac{2\hbar^2[Z_2(m,n_1,E_{m,n_1,s}) - Z_2(m,n_1,E_{m,n_1',s'})]}{\mu(E_{m,n_1,s} - E_{m,n_1',s'})}$$
$$\times \int_0^\rho g(m,n_1,E_{m,n_1,s};\eta) g(m,n_1',E_{m,n_1',s'};\eta)\frac{d\xi}{\eta}$$
$$\text{if} \quad E_{m,n_1,s} \neq E_{m,n_1',s'}. \tag{2.41}$$

Inserting (2.40) and (2.41) into (2.36), and noting that from (2.29) it follows that

$$Z_1(m, n_1, E_{m,n_1,s}) - Z_1(m, n_1', E_{m,n_1',s'}) \\ = -[Z_2(m, n_1, E_{m,n_1,s}) - Z_2(m, n_1', E_{m,n_1',s'})], \quad (2.42)$$

we obtain

$$\iiint \chi^*(m, n_1, E_{m,n_1,s}; x, y, z) \\ \times \chi(m, n_1', E_{m,n_1',s'}; x, y, z) dx\, dy\, dz = 0 \\ \text{if} \quad E_{m,n_1,s} \neq E_{m,n_1',s'}. \quad (2.43)$$

Since the energy eigenvalues $E_{m,n_1,s}$ are enumerable, while the number of possible values of ρ is *not* enumerable, we can choose ρ such that $E_{m,n_1,s} \neq E_{m,n_1',s'}$ for all quantum numbers unless $m = m', n_1 = n_1'$, and $s = s'$. We can also justify the possibility of choosing ρ in this way by noting that there is only a *finite*, although very large, number of quantum numbers that is relevant for the problem under consideration. The states for which $E_{m,n_1,s} = E_{m,n_1',s'}$ can thus be disregarded unless $n_1' = n_1$ and $s' = s$. Choosing ρ in the above-mentioned way, we obtain from (2.35) and (2.43)

$$\iiint \chi^*(m, n_1, E_{m,n_1,s}; x, y, z) \\ \times \chi(m', n_1', E_{m',n_1',s'}; x, y, z) dx\, dy\, dz = 0 \\ \text{unless} \quad m = m', \quad n_1 = n_1', \quad s = s'. \quad (2.44)$$

When $F \neq 0, m = m', n_1 = n_1, s = s'$ and ρ is sufficiently large, the first term on the right-hand side of (2.36) is negligible compared to the second term because of the magnitudes of the η-integrals. Hence we obtain approximately

$$\iiint \chi^*(m, n_1, E_{m,n_1,s}; x, y, z) \chi(m, n_1, E_{m,n_1,s}; x, y, z) dx\, dy\, dz \\ = [\Omega(m, n_1, E_{m,n_1,s})]^2 \int_0^\infty [f(m, n_1, E_{m,n_1,s}; \xi)]^2 \frac{d\xi}{4\xi} \\ \times \int_0^\rho [g(m, n_1, E_{m,n_1,s}; \xi)]^2 d\eta. \quad (2.45)$$

An alternative justfication of the approximation of (2.36) that leads to (2.45) will be given in the paragraph containing (5.20) and (5.21) in Chapter 5. If we normalize $f(m, n_1, E_{m,n_1,s}; \xi)$ and $g(m, n_1, E_{m,n_1,s}; \eta)$ such that

$$\int_0^\infty [f(m, n_1, E_{m,n_1,s}; \xi)]^2 \frac{d\xi}{4\xi} = 1, \tag{2.46}$$

$$[\Omega(m, n_1 E_{m,n_1,s})]^2 \int_0^\rho [g(m, n_1 E_{m,n_1,s}; \eta)]^2 d\eta = 1, \tag{2.47}$$

we obtain from (2.45)

$$\iiint \chi^*(m, n_1, E_{m,n_1,s}; x, y, z)\chi(m, n_1, E_{m,n_1,s}; x, y, z) dx\, dy\, dz = 1. \tag{2.48}$$

From (2.44) and (2.48) it follows that

$$\begin{aligned} \iiint \chi^*(m, n_1, E_{m,n_1,s}; x, y, z)\chi(m', n_1', E_{m',n_1',s'}; x, y, z) dx\, dy\, dz \\ = \delta_{m,m'}\delta_{n_1,n_1'}\delta_{s,s'}. \end{aligned} \tag{2.49}$$

Chapter 3

Development in Time of the Probability Amplitude for a Decaying State

The time-dependent wave function $\psi(x, y, z; t)$ for the internal motion is expanded as a superposition of solutions of the form (2.15) in the following way

$$\psi(x, y, z; t) = \sum_{m,n_1,s} C(m, n_1, E_{m,n_1,s})\chi(m, n_1, E_{m,n_1,s}; x, y, z) \times \exp(-iE_{m,n_1,s}t/\hbar). \tag{3.1}$$

From (2.49) and (3.1) it follows that

$$\iiint \psi^*(x, y, z; t)\psi(x, y, z; t)dx\, dy\, dz = \sum_{m,n_1,s} |C(m, n_1, E_{m,n_1,s})|^2 \tag{3.2}$$

and hence

$$\iiint \psi^*(x, y, z; t)\psi(x, y, z; t)dx\, dy\, dz = 1 \tag{3.3}$$

if

$$\sum_{m,n_1,s} |C(m, n_1, E_{m,n_1,s})|^2 = 1. \tag{3.4}$$

The coefficients $C(m, n_1, E_{m,n_1,s})$ are determined from the requirement that at the time $t = 0$ the wave function $\psi(x, y, z, t)$ is equal to a given wave function $\psi(x, y, z; 0)$. Using (3.1) and (2.49), one therefore finds that

$$C(m, n_1, E_{m,n_1,s}) = \iiint \chi^*(m, n_1, E_{m,n_1,s}; x, y, z) \times \psi(x, y, z; 0)dx\, dy\, dz. \tag{3.5}$$

The probability amplitude $p(t)$ that the electron is still in the initial state $\psi(x, y, z; 0)$ at the time $t(\geq 0)$ is by definition

$$p(t) = \iiint \psi^*(x, y, z; 0)\psi(x, y, z; t)dx\, dy\, dz, \tag{3.6}$$

and the probability that the electron is still in the initial state $\psi(x, y, z; 0)$ at the time $t(\geq 0)$ is $|p(t)|^2$. Inserting the expansion (3.1) for $\psi(x, y, z; t)$ and the corresponding expansion for $\psi(x, y, z; 0)$ into (3.6) and using (2.49), we obtain the formula

$$p(t) = \sum_{m,n_1,s} |C(m, n_1, E_{m,n_1,s})|^2 \exp\left(-\frac{iE_{m,n_1,s}t}{\hbar}\right), \tag{3.7}$$

which is analogous to the Fock–Krylov theorem; see Krylov and Fock (1947) and Drukarev, Fröman and Fröman (1979).

With the aid of (2.33) we can write (3.1) as

$$\begin{aligned}\psi(x, y, z; t) = \sum_{m,n_1,s} & C(m, n_1, E_{m,n_1,s})\Omega(m, n_1, E_{m,n_1,s}) \\ & \times \overline{\chi}(m, n_1, E_{m,n_1,s}; x, y, z) \exp\left(-\frac{iE_{m,n_1,s}t}{\hbar}\right)\end{aligned} \tag{3.8}$$

and (3.5) as

$$\begin{aligned}&C(m, n_1, E_{m,n_1,s}) \\ &\quad = \Omega(m, n_1, E_{m,n_1,s}) \iiint \overline{\chi}^*(m, n_1, E_{m,n_1,s}; x, y, z) \\ &\quad \times \psi(x, y, z; 0)dx\, dy\, dz,\end{aligned} \tag{3.9}$$

Ω being real.

Assuming that ρ is sufficiently large, in order that, according to (5.25) in Chapter 5, the difference $\Delta E = E_{m,n_1,s+1} - E_{m,n_1,s}$ between two neighboring energy levels with the same quantum numbers m and n_1 be so small that the sum with respect to s can be replaced by an integral over E, we can write (3.8) as

$$\begin{aligned}\psi(x, y, z; t) = \sum_{m=-\infty}^{\infty} \sum_{n_1=0}^{\infty} \int & \overline{C}(m, n_1, E) \\ & \times \overline{\chi}(m, n_1, E; x, y, z) \exp\left(-\frac{iEt}{\hbar}\right) dE,\end{aligned} \tag{3.10}$$

where $\overline{\chi}(m, n_1, E; x, y, z)$ is given by (2.34), with $E_{m,n_1,s}$ replaced by E, and

$$\overline{C}(m, n_1, E) = \frac{C(m, n_1, E)\Omega(m, n_1, E)}{\Delta E}. \tag{3.11}$$

With the use of (3.9), with $E_{m,n_1,s}$ replaced by E, we can write (3.11) as

$$\overline{C}(m, n_1, E) = \frac{[\Omega(m, n_1, E)]^2}{\Delta E} \iiint \overline{\chi}^*(m, n_1, E; x, y, z) \times \psi(x, y, z; 0) dx\, dy\, dz. \tag{3.12}$$

By similarly replacing the sum with respect to s by an integral over E, one can with the use of (3.11) write (3.4) as

$$\sum_{m=-\infty}^{\infty} \sum_{n_1=0}^{\infty} \int \frac{|C(m, n_1, E)|^2}{\Delta E} dE = \sum_{m=-\infty}^{\infty} \sum_{n_1=0}^{\infty} \int \frac{\Delta E}{[\Omega(m, n_1, E)]^2} |\overline{C}(m, n_1, E)|^2 dE = 1 \tag{3.13}$$

and (3.7) as

$$p(t) = \sum_{m=-\infty}^{\infty} \sum_{n_1=0}^{\infty} \int \frac{|C(m, n_1, E)|^2}{\Delta E} \exp\left(-\frac{iEt}{\hbar}\right) dE = \sum_{m=-\infty}^{\infty} \sum_{n_1=0}^{\infty} \int \frac{\Delta E}{[\Omega(m, n_1, E)]^2} |\overline{C}(m, n_1, E)|^2 \exp\left(-\frac{iEt}{\hbar}\right) dE. \tag{3.14}$$

Chapter 4

Phase-Integral Method

Since the treatment in Chapter 5 is based on phase-integral formulas that are scattered in different publications, we collect in the present chapter background material that is necessary for reading Chapter 5.

The phase-integral method for solving differential equations of the type

$$\frac{d^2\psi}{dz^2} + R(z)\psi = 0, \tag{4.1}$$

where $R(z)$ is an unspecified analytic function of the complex variable z, involves the following items:

(i) Arbitrary-order phase-integral approximation generated from an unspecified base function $Q(z)$ as described in Chapter 1 of Fröman and Fröman (1996) and in Fröman and Fröman (2002); see also Dammert and P. O. Fröman (1980).

(ii) Method for solving connection problems developed by Fröman and Fröman (1965), generalized to apply to the phase-integral approximation referred to in item (i).

(iii) Supplementary quantities, expressed analytically in terms of phase-integrals. An example is the quantity $\tilde{\phi}$, which is of decisive importance, when two generalized classical turning points of a potential barrier lie close to each other; see Section 4.3.

We shall first briefly describe the phase-integral approximation referred to in item (i). Then we collect connection formulas pertaining to a single transition point [first-order zero or first-order pole of $Q^2(z)$] and to a real potential barrier, which can be derived by means of the method mentioned in item (ii) combined with comparison equation technique for obtaining the supplementary quantity $\tilde{\phi}$ mentioned in item (iii) and appearing in the connection formula for a real potential barrier. Finally we present quantization conditions for single-well potentials, which can be derived by means of the connection formulas pertaining to a single transition point.

4.1 Phase-integral approximation generated from an unspecified base function

For a detailed description of this approximation we refer to Chapter 1 in Fröman and Fröman (1996) and to Fröman and Fröman (2002). A brief description is given below.

We introduce into (4.1) a "small" bookkeeping parameter λ that will finally be put equal to unity. Thus we get the auxiliary differential equation

$$\frac{d^2\psi}{dz^2} + \left[\frac{Q^2(z)}{\lambda^2} + R(z) - Q^2(z)\right]\psi = 0, \tag{4.2}$$

which goes over into (4.1) when $\lambda = 1$. The function $Q(z)$ is the unspecified base function from which the phase-integral approximation is generated. This function is often chosen to be equal to $R^{1/2}(z)$, but in many physical problems it is important to use the possibility of choosing $Q(z)$ differently in order to achieve the result that the phase-integral approximation be valid close to certain exceptional points [e.g., the origin in connection with the radial Schrödinger equation, and the poles of $\tilde{Q}^2(\xi)$ and $Q^2(\eta)$ at $\xi = 0$ and $\eta = 0$ in the Stark effect problem treated in this book; see Eqs. (5.1) and (5.2)], where the approximation would fail, if $Q(z)$ were chosen to be equal to $R^{1/2}(z)$. The function $Q(z)$ is in general chosen such that it is approximately equal to $R^{1/2}(z)$ except possibly in the neighborhood of the exceptional points.

The auxiliary differential equation (4.2) has two linearly independent solutions $f_1(z)$ and $f_2(z)$ of the form

$$f_1(z) = q^{-1/2}(z)\exp[+iw(z)], \tag{4.3a}$$

$$f_2(z) = q^{-1/2}(z)\exp[-iw(z)], \tag{4.3b}$$

where

$$w(z) = \int^z q(z)dz. \tag{4.4}$$

Inserting (4.3a) or (4.3b) for ψ into (4.2), we obtain

$$q^{+1/2}\frac{d^2q^{-1/2}}{dz^2} - q^2 + \frac{Q^2(z)}{\lambda^2} + R(z) - Q^2(z) = 0. \tag{4.5}$$

Introducing instead of z the variable

$$\zeta = \int^z Q(z)dz, \tag{4.6}$$

we can write (4.5) in the form

$$1 - \left(\frac{q\lambda}{Q(z)}\right)^2 + \varepsilon_0\lambda^2 + \left(\frac{q\lambda}{Q(z)}\right)^{+1/2}\frac{d^2}{d\zeta^2}\left(\frac{q\lambda}{Q(z)}\right)^{-1/2}\lambda^2 = 0, \tag{4.7}$$

where

$$\varepsilon_0 = Q^{-3/2}(z)\frac{d^2Q^{-1/2}(z)}{dz^2} + \frac{R(z)}{Q^2(z)} - 1. \tag{4.8}$$

To obtain a formal solution of (4.7), we put

$$\frac{q\lambda}{Q(z)} = \sum_{n=0}^{\infty} Y_{2n}\lambda^{2n}, \tag{4.9}$$

where Y_0 is assumed to be different from zero, and Y_{2n} ($n = 0, 1, 2, \ldots$) are independent of λ. Inserting the expansion (4.9) into (4.7), expanding the left-hand side in powers of λ, and putting the coefficient of each power of λ equal to zero, we get $Y_0 = \pm 1$ and a recurrence formula, from which one can successively obtain the functions $Y_2, Y_4, Y_6, \ldots$, each one of which can be expressed in terms of ε_0, defined in (4.8), and derivatives of ε_0 with respect to ζ. Since we have both $+$ and $-$ in the exponents of (4.3a,b), it is no restriction to choose $Y_0 = 1$. The first few functions Y_{2n} are then

$$Y_0 = 1, \tag{4.10a}$$

$$Y_2 = \frac{1}{2}\varepsilon_0, \tag{4.10b}$$

$$Y_4 = -\frac{1}{8}\left({\varepsilon_0}^2 + \frac{d^2\varepsilon_0}{d\zeta^2}\right). \tag{4.10c}$$

The choice of the unspecified base function $Q(z)$ shows itself only in the expressions (4.6) and (4.8) for ζ and ε_0 which depend explicitly on $R(z)$ and $Q(z)$, while the functions Y_{2n}, which are expressed in terms of ε_0 and derivatives of ε_0 with respect to ζ, do not depend explicitly on $R(z)$ and the choice of the base function $Q(z)$. The expressions for the functions Y_{2n} can therefore be determined once and for all. We also remark that at the zeros and poles of $Q^2(z)$ the functions $Q(z)$ and $Q^{-1/2}(z)$ may have branch points, whereas the functions ε_0, Y_{2n} and $q(z)/Q(z)$ are single-valued. Truncating the infinite series in (4.9) at $n = N$, we obtain

$$q(z) = Q(z) \sum_{n=0}^{N} Y_{2n} \lambda^{2n-1}. \tag{4.11}$$

Inserting (4.11) into (4.3a,b) along with (4.4) and putting $\lambda = 1$, we get the phase-integral functions of the order $2N + 1$, generated from the base function $Q(z)$, which are approximate solutions of the differential equation (4.1). For $N > 0$ the function $q(z)$ has poles at the transition zeros, i.e., the zeros of $Q^2(z)$, and simple zeros in the neighborhood of each transition zero (N. Fröman 1970).

In the first order the phase-integral approximation is the same as the usual Carlini (JWKB) approximation[1] if $Q(z) = R^{1/2}(z)$, but in higher orders it differs in essential respects from that approximation of corresponding order; see Dammert and P. O. Fröman (1980) and Chapter 1 in Fröman and Fröman (1996). Although the phase-integral approximation generated from an unspecified base function is in higher order essentially different from the Carlini approximation, there are between these two approximations relations which we shall now discuss. According to (4.3a,b), (4.4) and (4.11) with $\lambda = 1$ the $(2N + 1)$th-order phase-integral approximation generated from the base function $Q(z)$ is

$$\psi = \frac{\exp\left[\pm i \int_{z_0}^{z} Q(z) \sum_{n=0}^{N} Y_{2n} dz\right]}{\left[Q(z) \sum_{n=0}^{N} Y_{2n}\right]^{1/2}}. \tag{4.12}$$

[1] As regards the motivation for the name Carlini approximation we refer to Fröman and Fröman (1985) or to Chapter 1 in Fröman and Fröman (2002).

When one derives the Carlini (JWKB) approximation one introduces a factor $1/\lambda^2$ in front of $R(z)$ in the differential equation (4.1) and puts

$$\psi = \exp\left[i\int_{z_0}^{z}\sum_{n=0}^{\infty} y_{2n}(z)\lambda^{n-1}\,dz\right]. \tag{4.13}$$

One obtains (after λ has been put equal to unity) in the first order

$$\psi = \frac{\exp\left[\pm i\int_{z_0}^{z} R^{1/2}(z)\,dz\right]}{R^{1/4}(z)}, \tag{4.14a}$$

in the third order

$$\psi = \frac{\exp\left[\pm i\int_{z_0}^{z} R^{1/2}(z)(1+Y_2)\,dz\right]}{R^{1/4}(z)\exp\left(\frac{Y_2}{2}\right)}, \tag{4.14b}$$

and in the fifth order

$$\psi = \frac{\exp\left[\pm i\int_{z_0}^{z} R^{1/2}(z)(1+Y_2+Y_4)\,dz\right]}{R^{1/4}(z)\exp\left[\frac{Y_2}{2}+\left(\frac{Y_4}{2}-\frac{Y_2^2}{4}\right)\right]}, \tag{4.14c}$$

where Y_2 and Y_4 are given by (4.10b) and (4.10c) along with (4.6) and (4.8) with $Q^2(z) = R(z)$. It is seen that for the phase-integral approximation there is in every order a simple connection between phase and amplitude, while for the higher orders of the Carlini (JWKB) approximation the expression for the amplitude is complicated. When the base function $Q(z)$ is chosen to be equal to $R^{1/2}(z)$ the phase in a classically allowed region is the same for both approximations. It should also be remarked that when one determines the functions Y_{2n} with $n > 0$ by means of the recurrence formula for the phase-integral approximation, one obtains directly the simple expressions (4.10b,c), but when one determines these functions by means of the usual recurrence formula for the Carlini (JWKB) approximation, one obtains for $n > 0$ complicated expressions, the simplification of which to the form (4.10b,c) requires rather complicated calculations.

The criterion for the determination of the base function is that ε_0 be small compared to unity in the region of the complex z-plane relevant for the problem under consideration. As an example of how this was done in a situation where the condition for the validity of the semi-classical approximation is not fulfilled, and hence the choice $Q^2(z) = R(z)$ is not very useful, we refer to N. Fröman (1979) and in particular to Eqs. (14) and (58) there. With her appropriate choice of $Q^2(z)$, which is significantly different from $R(z)$ in the whole relevant region of the complex z-plane, she obtained very accurate results; see Table 4 in N. Fröman (1979).

It is an essential advantage of the phase-integral approximation generated from an unspecified base function versus the Carlini (JWKB) approximation that the former approximation contains the unspecified base function $Q(z)$, which one can take advantage of in several ways. The criterion for the determination of the base function mentioned in the previous paragraph does not determine $Q(z)$ uniquely. It turns out that, within certain limits, the results are not very sensitive to the choice of $Q(z)$, when the approximation is used in higher orders. However, with a convenient choice of $Q(z)$ already the first-order approximation can be very good. On the other hand, an inconvenient, but possible, choice of $Q(z)$ introduces in the first-order approximation an unnecessarily large error that is, however, in general corrected already in the third-order approximation. In many important cases the function $Q^2(z)$ can be chosen to be identical to $R(z)$. In other important cases, for instance, when one wants to include the immediate neighborhood of a first- or second-order pole of $R(z)$ in the region of validity of the phase-integral approximation, the function $Q^2(z)$ is in general chosen to be approximately equal to $R(z)$ except in the neighborhood of the pole.

The freedom that one has in the choice of the base function $Q(z)$ will now be illuminated in a concrete way. For a radial Schrödinger equation the usual choice of $Q^2(z)$ is

$$Q^2(z) = R(z) - \frac{1}{(4z^2)}. \tag{4.15a}$$

However, the replacement of (4.15a) by

$$Q^2(z) = R(z) - \frac{1}{(4z^2)} - \frac{\text{const}}{z}, \tag{4.15b}$$

where the coefficient of $1/z$ should be comparatively small, does not destroy the great accuracy of the results usually obtained with the phase-integral approximation in higher orders. There is thus a whole set of base functions that may be used, and there are various ways in which one can take advantage of this non-uniqueness to make the choice of the base function well adapted to the particular problem under consideration. For instance, by adapting the choice of $Q^2(z)$ to the analytical form of $R(z)$ one can sometimes achieve the result that the integrals occurring in the phase-integral approximation can be evaluated analytically. To give an example we assume that $R(z)$ contains only $\exp(z)$ but not z itself. In this case it is convenient to replace the choice (4.15b) by the choice

$$Q^2(z) = R(z) - \frac{1}{4(e^z - 1)^2} - \frac{\text{const}}{e^z - 1}. \tag{4.15c}$$

By a convenient choice of $Q^2(z)$, for instance a convenient choice of the unspecified coefficient in (4.15b) or (4.15c), one can sometimes attain the result that, for example, eigenvalues or phase-shifts are obtained exactly for some particular parameter value in every order of the phase-integral approximation. By making this exactness fulfilled in the limit of a parameter value, for which the phase-integral result without this adaptation would not be good, one can actually extend the region of validity of the phase-integral treatment; see pages 16 and 17 in Fröman, Fröman and Larsson (1994). When the differential equation contains one or more parameters, the accurate calculation of the wave function may require different choices of the base function $Q(z)$ for different ranges of the parameter values. To illustrate this fact we consider a radial Schrödinger equation. For sufficiently large values of the angular momentum quantum number l we obtain an accurate phase-integral approximation (valid also close to $z = 0$) if we choose $Q^2(z)$ according to (4.15a), (4.15b) or (4.15c). If the value of l is too small, the phase-integral approximation with this choice of $Q^2(z)$ is not good. It can be considerably improved (except close to $z = 0$), when the absolute value of the coefficient of $1/z$ in $R(z)$ is sufficiently large, if one instead chooses

$$Q^2(z) = R(z) + \frac{l(l+1)}{z^2}. \tag{4.15d}$$

The corresponding phase-integral approximation is not valid close to $z = 0$, but the wave function that is regular and tends to z^{l+1}, when z is a dimensionless variable that tends to zero, can be obtained sufficiently far away from $z = 0$ by means of the connection formula that will be presented in Subsection 4.2.2 of this chapter. The presence of the unspecified base function $Q(z)$ in the phase-integral approximation is thus very important from several points of view.

When the first-order approximation is used, it is often convenient to choose the constant lower limit of integration in the definition (4.4) of $w(z)$ to be a zero or a first-order pole of $Q^2(z)$. This is, however, in general not possible when a higher-order approximation is used, since the integral in (4.4) would then in general be divergent. If the lower limit of integration in (4.4) is an odd-order zero or an odd-order pole of $Q^2(z)$, it is possible and convenient to replace the definition (4.4) of $w(z)$ by the definition (N. Fröman 1966b, 1966c, 1970)

$$w(z) = \frac{1}{2} \int_{\Gamma_t(z)} q(z) dz, \tag{4.16}$$

where t is the odd-order zero or odd-order pole in question, and $\Gamma_t(z)$ is a path of integration that starts at the point corresponding to z on a Riemann sheet adjacent to the complex z-plane under consideration, encircles t in the positive or in the negative direction and ends at z. It is immaterial for the value of the integral in (4.16) whether the path of integration encircles t in the positive or in the negative direction, but the terminal point must be the point z in the complex z-plane under consideration. For the first-order approximation the definition (4.4), with the lower limit of integration equal to t, and the definition (4.16) are identical.

It is useful to introduce a short-hand notation for the integral on the right-hand side of (4.16) by the definition

$$\int_{(t)}^{z} q(z) dz = \frac{1}{2} \int_{\Gamma_t(z)} q(z) dz. \tag{4.17}$$

For the first order of the phase-integral approximation one can replace (t) by t on the left-hand side of (4.17) and thus get an ordinary integral from t to z instead of half of the integral along the contour $\Gamma_t(z)$. In analogy to (4.17) one defines a short-hand notation for an integral in which the upper limit of integration is an odd-order zero or an odd-order pole of $Q^2(z)$. When one has two transition points of that kind as limits of integration, one requires that the contours of integration pertaining to the lower and upper limits of integration are encircled in the same direction. The definition of the short-hand notation with both limits within parentheses implies then that the integral is equal to half of the integral along a closed loop enclosing both transition points. The simplified notation on the left-hand side of (4.17) for the integral on the right-hand side of (4.17) was introduced by Fröman, Fröman and Lundborg (1988), pages 160 and 161. It makes it possible to use, for an arbitrary order of the phase-integral approximation, a similar simple notation and almost the same simple language (although in a generalized sense) as for the first order of the phase-integral approximation. One thus achieves a great formal and practical simplification in the treatment of concrete problems, when an arbitrary order of the phase-integral approximation is used.

We remark that the notations used above differ from the notations in the original papers published up to the middle of the 1980s in the respect that $Q^2(z)$ and $Q^2_{\text{mod}}(z)$ in those papers correspond in later publications, and thus in the present chapter, to $R(z)$ and $Q^2(z)$, respectively.

4.2 Connection formulas associated with a single transition point

4.2.1 *Connection formulas pertaining to a first-order transition zero on the real axis*

The phase-integral formulas in this subsection are valid when $R(z)$ and $Q^2(z)$ are real on the real z-axis. For the first order of the phase-integral approximation Fröman and Fröman (1965) presented rigorous derivations of these connection formulas. Before the phase-integral approximation generated from an unspecified base function had been introduced, N. Fröman (1970) derived arbitrary-order connection formulas associated with a first-order transition zero for the particular phase-integral approximation of arbitrary order corresponding to $Q^2(z) = R(z)$. After the phase-integral approximation generated from an unspecified base function had been introduced, it turned out that these connection formulas remain valid also when

$Q^2(z) \neq R(z)$. Below we shall present these general connection formulas.

As already mentioned, the functions $R(z)$ and $Q^2(z)$ are assumed to be real on the real z-axis (the x-axis). We assume that on this axis there is a generalized classical turning point t, i.e., a simple zero of $Q^2(z)$. In a generalized sense there is then on one side of t a classically allowed region, i.e., a region where $Q^2(x) > 0$, and on the other side of t a classically forbidden region, i.e., a region where $Q^2(x) < 0$. Defining

$$w(x) = \int_{(t)}^{x} q(z)dz, \tag{4.18}$$

we can write the connection formula for tracing a phase-integral solution of the differential equation (4.1) from the classically forbidden to the classically allowed region as

$$\begin{aligned} &|q^{-1/2}(x)| \exp[-|w(x)|] + C|q^{-1/2}(x)| \exp[|w(x)|] \\ &\quad \to 2|q^{-1/2}(x)| \cos\left[|w(x)| - \frac{\pi}{4}\right]; \end{aligned} \tag{4.19}$$

it is valid provided that the condition

$$C \exp\{|w(x)|\} \lesssim \exp\{-|w(x)|\} \tag{4.20}$$

is fulfilled at the point from which one makes the connection. A numerical study of the accuracy and the properties of the connection formula (4.19) with $C = 0$ has been published by N. Fröman and W. Mrazek (1977).

The connection formula for tracing a phase-integral solution of the differential equation (4.1) from the classically allowed to the classically forbidden region is

$$\begin{aligned} &A|q^{-1/2}(x)| \exp\left\{i\left[|w(x)| + \frac{\pi}{4}\right]\right\} \\ &\quad + B|q^{-1/2}(x)| \exp\left\{-i\left[|w(x)| + \frac{\pi}{4}\right]\right\} \\ &\quad \to (A+B)|q^{-1/2}(x)| \exp[|w(x)|], \end{aligned} \tag{4.21}$$

where A and B are constants, which are arbitrary, except for the requirement that $|A+B|/(|A|+|B|)$ must not be too close to zero.

As a consequence of (4.21) one obtains the connection formula

$$|q^{-1/2}(x)|\cos\left[|w(x)|+\delta-\frac{\pi}{4}\right]\to\sin\delta|q^{-1/2}(x)|\exp[|w(x)|], \quad (4.22)$$

where δ is a real phase constant that must not be too close to a multiple of π. We emphasize the one-directional validity of the connection formulas (4.19), (4.21) and (4.22), which means that the tracing of a solution must always be made in the direction of the arrow. This property of the connection formulas has been thoroughly investigated and even illustrated numerically by N. Fröman (1966a) for the first order of the Carlini (JWKB) approximation. The whole discussion in that paper applies in principle also to the connection formulas for the higher orders of the phase-integral approximation generated from an unspecified base function. The one-directional validity of the connection formula (4.22) is obvious, since a change of the phase δ on the left-hand side of (4.22) causes a change of the amplitude on the right-hand side of (4.22), while a change of the amplitude of the wave function in the classically forbidden region cannot cause a change of the phase of the wave function in the classically allowed region.

The arbitrary-order connection formulas (4.19), (4.21) and (4.22) can in many cases be used for obtaining very accurate solutions of physical problems, when the turning points are well separated, and there are no other transition points near the real axis in the region of the complex z-plane of interest. Within their range of applicability, these connection formulas are very useful because of their simplicity and the great ease with which they can be used. They have been discussed by Fröman and Fröman (2002); see Sections 3.10–3.13 and 3.20 there.

4.2.2 *Connection formula pertaining to a first-order transition pole at the origin*

Now we assume that in a certain region of the complex z-plane around a first-order transition pole at the origin, i.e., a first-order pole of $Q^2(z)$ at the origin, we have

$$R(z)=-\frac{l(l+1)}{z^2}+\frac{B}{z}+\begin{matrix}\text{a function of } z \text{ that is regular}\\ \text{in a region around the origin}\end{matrix}, \quad (4.23)$$

where $2l + 1$ is a non-negative integer, and

$$Q^2(z) = \frac{\bar{B}}{z} + \begin{array}{l}\text{a function of } z \text{ that is regular}\\ \text{in a region around the origin}\end{array}. \tag{4.24}$$

We also assume that the absolute values of B and $\bar{B}$ are sufficiently large, while the absolute value of $B - \bar{B}$ and the absolute value of the difference between the two regular functions in (4.23) and (4.24) are sufficiently small. There is one particular curve on which $w(z)$, defined as

$$w(z) = \int_{(0)}^{z} q(z)dz, \tag{4.25}$$

is real. For the first order of the phase-integral approximation this is the anti-Stokes line that emerges from the origin. Therefore we use, also for a higher order of the phase-integral approximation, the terminology "the anti-Stokes line that emerges from the origin" in a generalized sense to denote the line on which $w(z)$ in (4.25) is real. For the first-order approximation Fröman and Fröman (1965) obtained a phase-integral formula [their Eq. (7.28)], valid sufficiently far away from the origin on the anti-Stokes line that emerges from the origin, for the particular solution $\psi(z)$ of the differential equation (4.1) that fulfills the condition

$$\lim_{z \to 0} \frac{\psi(z)}{z^{l+1}} = 1, \tag{4.26}$$

where z is dimensionless. That formula can be generalized to be valid for an arbitrary order of the phase-integral approximation generated from an unspecified base function and under less restrictive assumptions than in the original derivation. It can then be formulated as follows. On the lip of the anti-Stokes line emerging from the origin, where $w(z) = |w(z)|$, the solution of the differential equation (4.1) that fulfils the condition (4.26) is, sufficiently far away from the origin, given by the phase-integral formula

$$\psi(z) = (\pi c)^{-1/2} q^{-1/2}(z) \cos\left[w(z) - \left(l + \frac{3}{4}\right)\pi\right], \tag{4.27}$$

where c is the residue of $[\psi(z)]^{-2}$ at the origin, and the sign of $(\pi c)^{-1/2}$ has to be chosen appropriately; c is thus determined by the expansion of $\psi(z)$ in powers of z. For the special case that $l = 0$

one finds that $c = B$. When $R(z)$ and $Q^2(z)$ are real on the real axis (the x-axis), and z is a point x in the classically allowed region $[Q^2(x) > 0]$ adjacent to the origin, (4.27) can be written as

$$\psi(x) = (\pi c)^{-1/2} |q^{-1/2}(x)| \cos \left[\left| \int_{(0)}^{x} q(z)\,dz \right| - \left(l + \frac{3}{4} \right) \pi \right]. \tag{4.28}$$

If, in particular, $l = -1/2$ this formula particularizes to

$$\psi(x) = |q^{-1/2}(x)| \cos \left[\left| \int_{(0)}^{x} q(z)\,dz \right| - \frac{\pi}{4} \right] \tag{4.29}$$

except for a constant factor.

4.3 Connection formula for a real, smooth, single-hump potential barrier

We denote by t' and t'' the two relevant zeros of $Q^2(z)$, i.e., the two generalized classical turning points in the sub-barrier case ($t' < t''$) and the two complex conjugate transition zeros in the super-barrier case ($\operatorname{Im} t' \leq 0, \operatorname{Im} t'' \geq 0$), and we let x' be a point in the classically allowed region of the real z-axis to the left of the barrier and x'' a point in the classically allowed region of the real z-axis to the right of the barrier. The points x' and x'' must not lie too close to the points t' and t''. Using the short-hand notation defined in (4.17), we obtain from Eqs. (2.5.4a) and (2.5.4b) in Fröman and Fröman (2002)

$$\psi(x') = A' |q^{-1/2}(x')| \exp \left[+i \left| \operatorname{Re} \int_{(t')}^{x'} q(z)\,dz \right| \right] + B' |q^{-1/2}(x')| \exp \left[-i \left| \operatorname{Re} \int_{(t')}^{x'} q(z)\,dz \right| \right] \tag{4.30a}$$

and

$$\psi(x'') = A'' |q^{-1/2}(x'')| \exp \left[+i \left| \operatorname{Re} \int_{(t'')}^{x''} q(z)\,dz \right| \right] + B'' |q^{-1/2}(x'')| \exp \left[-i \left| \operatorname{Re} \int_{(t'')}^{x''} q(z)\,dz \right| \right], \tag{4.30b}$$

where according to Eqs. (2.5.5) and (2.5.6a) in Fröman and Fröman (2002)

$$\begin{pmatrix} A'' \\ B'' \end{pmatrix} = \tilde{\mathbf{M}} \begin{pmatrix} A' \\ B' \end{pmatrix} \tag{4.31}$$

with

$$\tilde{\mathbf{M}} = \begin{pmatrix} \theta \exp\left[-i\left(\frac{\pi}{2}+\vartheta\right)\right] & (\theta^2+1)^{1/2}\exp(+i\tilde{\phi}) \\ (\theta^2+1)^{1/2}\exp(-i\tilde{\phi}) & \theta\exp\left[+i\left(\frac{\pi}{2}+\vartheta\right)\right] \end{pmatrix}. \tag{4.32}$$

When the transition points that are *not* associated with the barrier lie sufficiently far away from t' and t'', one has according to Eqs. (2.5.10a) and (2.5.2) in Fröman and Fröman (2002) for the quantity θ in (4.32) the formula

$$\theta \approx \exp(K) \tag{4.33}$$

with

$$K = \frac{1}{2i}\int_\Lambda q(z)dz, \tag{4.34}$$

where Λ is a closed contour of integration encircling both t' and t'', but no other transition point, with the integration performed in the direction that in the *first-order approximation* yields $K > 0$ for energies below the top of the barrier and $K < 0$ for energies above the top of the barrier. If higher-order approximations are used, the quantity K may become negative also for energies below (but not too far from) the top of the barrier; see Table 1 in N. Fröman (1980). When the transition points that are *not* associated with the barrier lie sufficiently far away from t' and t'', one has according to Eq. (2.5.10b) in Fröman and Fröman (2002) for the quantity ϑ in (4.32) the formula

$$\vartheta \approx 0. \tag{4.35}$$

The quantity $\tilde{\phi}$ in (4.32) will be discussed later. From (4.32) we obtain

$$\tilde{\mathbf{M}}^{-1} = \begin{pmatrix} \theta \exp\left[-i\left(\frac{\pi}{2}-\vartheta\right)\right] & (\theta^2+1)^{1/2}\exp(+i\tilde{\phi}) \\ (\theta^2+1)^{1/2}\exp(-i\tilde{\phi}) & \theta\exp\left[+i\left(\frac{\pi}{2}-\vartheta\right)\right] \end{pmatrix}. \tag{4.36}$$

It is seen that one obtains $\tilde{\mathbf{M}}^{-1}$ from $\mathbf{M}$ by replacing ϑ by $-\vartheta$. We emphasize that, except for (4.33) and (4.35), the above formulas are in principle exact, provided that one knows the quantities θ, ϑ and $\tilde{\phi}$, which depend slightly on x' and x''. Furthermore, if one knows θ, ϑ and $\tilde{\phi}$, the two transition zeros associated with the potential barrier need not lie very far away from transition points that are not associated with the barrier. However, when one introduces for θ, ϑ and $\tilde{\phi}$ the approximate expressions that for θ and ϑ are given in (4.33) and (4.35) and that for $\tilde{\phi}$ will be given in Subsection 4.3.2, the barrier is assumed to lie far away from all transition points that are not associated with the barrier.

When A' and B' are given constants, associated with a wave function that is given at the point x', the coefficients A'' and B'', which are obtained from (4.31) along with (4.32), depend slightly on x' and x'' via the quantities θ, ϑ and $\tilde{\phi}$, but one obtains the derivatives of $\psi(x')$ and $\psi(x'')$ from (4.30a,b) by considering A', B', A'' and B'' *formally* as constants.

4.3.1 *Wave function given as a standing wave*

The case when the wave function is given as a standing wave on one side of the barrier requires a detailed treatment, since the resonance phenomenon may occur. Putting in (4.30a)

$$A' = \frac{1}{2}\Omega' \exp\left[i\left(\delta' - \frac{\pi}{4}\right)\right], \tag{4.37a}$$

$$B' = \frac{1}{2}\Omega' \exp\left[-i\left(\delta' - \frac{\pi}{4}\right)\right], \tag{4.37b}$$

where Ω' is an arbitrary positive amplitude, and δ' is an arbitrary real phase, we get

$$\psi(x') = \Omega' |q^{-1/2}(x')| \cos\left[\left|\mathrm{Re}\int_{(t')}^{x'} q(z)dz\right| + \delta' - \frac{\pi}{4}\right], \tag{4.38a}$$

and putting in (4.30b)

$$A'' = \frac{1}{2}\Omega'' \exp\left[i\left(\delta'' - \frac{\pi}{4}\right)\right], \tag{4.39a}$$

$$B'' = \frac{1}{2}\Omega'' \exp\left[-i\left(\delta'' - \frac{\pi}{4}\right)\right], \tag{4.39b}$$

where Ω'' is a positive amplitude, and δ'' is a real phase, we obtain

$$\psi(x'') = \Omega''|q^{-1/2}(x'')| \cos\left[\left|\mathrm{Re}\int_{(t'')}^{x''} q(z)\,dz\right| + \delta'' - \frac{\pi}{4}\right]. \quad (4.38b)$$

To be able to discuss resonance phenomena we need exact or very accurate formulas admitting of a detailed analysis of how δ'' and Ω'' depend on δ' and Ω'. Inserting (4.37a,b) and (4.39a,b) into (4.31) along with (4.32), and writing the resulting equation in a convenient form, we obtain

$$\begin{aligned}\Omega'' \exp\left[i\left(\delta'' - \frac{\tilde{\phi}}{2} + \frac{\vartheta}{2}\right)\right] &= \Omega'\left\{(\theta^2+1)^{1/2}\exp\left[+i\left(\frac{\pi}{2} + \frac{\tilde{\phi}}{2} + \frac{\vartheta}{2} - \delta'\right)\right]\right. \\ &\left.+ \theta\exp\left[-i\left(\frac{\pi}{2} + \frac{\tilde{\phi}}{2} + \frac{\vartheta}{2} - \delta'\right)\right]\right\} \quad (4.40)\end{aligned}$$

and the complex conjugate of (4.40). Separating (4.40) into real and imaginary parts, we get

$$\Omega''\cos\left(\delta'' - \frac{\tilde{\phi}}{2} + \frac{\vartheta}{2}\right) = \Omega'[(\theta^2+1)^{1/2} + \theta]\cos\left(\frac{\pi}{2} + \frac{\tilde{\phi}}{2} + \frac{\vartheta}{2} - \delta'\right), \quad (4.41a)$$

$$\Omega''\sin\left(\delta'' - \frac{\tilde{\phi}}{2} + \frac{\vartheta}{2}\right) = \Omega'[(\theta^2+1)^{1/2} - \theta]\sin\left(\frac{\pi}{2} + \frac{\tilde{\phi}}{2} + \frac{\vartheta}{2} - \delta'\right). \quad (4.41b)$$

According to (4.41a,b) the angle $\delta'' - \tilde{\phi}/2 + \vartheta/2$ lies in the same quadrant as the angle $\pi/2 + \tilde{\phi}/2 + \vartheta/2 - \delta'$ (mod 2π). From (4.41a,b) we get

$$\delta'' = \arctan\left[\frac{(\theta^2+1)^{1/2} - \theta}{(\theta^2+1)^{1/2} + \theta}\tan\left(\frac{\pi}{2} + \frac{\tilde{\phi}}{2} + \frac{\vartheta}{2} - \delta'\right)\right] + \frac{\tilde{\phi}}{2} + \frac{\vartheta}{2}, \quad (4.42a)$$

$$\Omega'' = \Omega'\left\{[(\theta^2+1)^{1/2} - \theta]^2 + 4\theta(\theta^2+1)^{1/2}\cos^2\left(\frac{\pi}{2} + \frac{\tilde{\phi}}{2} + \frac{\vartheta}{2} - \delta'\right)\right\}^{1/2}, \quad (4.42b)$$

where the branch of arctan is to be chosen such that δ'' is a continuous function of δ', with $\delta'' - \tilde{\phi}/2 + \vartheta/2$ lying in the same quadrant as

$\pi/2 + \tilde{\phi}/2 + \vartheta/2 - \delta'$ (mod 2π). Alternatively we can write (4.42a) and (4.42) as

$$\delta'' = \arctan\left[\frac{u-1}{u+1}\tan\left(\frac{\pi}{2} - v\right)\right] + \frac{\tilde{\phi}}{2} - \frac{\vartheta}{2}, \tag{4.43a}$$

$$\begin{aligned} \Omega'' &= \Omega'\left[\frac{u-1}{u+1} + \frac{4u}{u^2-1}\sin^2 v\right]^{1/2} \\ &= \Omega'\left[\frac{u+1}{u-1} - \frac{4u}{(u^2-1)(1+\tan^2 v)}\right]^{1/2}, \end{aligned} \tag{4.43b}$$

where

$$u = (1 + 1/\theta^2)^{1/2}, \tag{4.44a}$$

$$v = \delta' - \frac{\tilde{\phi}}{2} - \frac{\vartheta}{2}. \tag{4.44b}$$

From (4.43b) it follows that

$$\left(\frac{u-1}{u+1}\right)^{1/2} \leq \frac{\Omega''}{\Omega'} \leq \left(\frac{u+1}{u-1}\right)^{1/2}, \tag{4.45}$$

where the equality sign to the left is valid when v is an integer multiple of π, and the equality sign to the right is valid when $v - \pi/2$ is an integer multiple of π.

4.3.2 *Supplementary quantity $\tilde{\phi}$*

The quantity $\tilde{\phi}$ is particularly important when the energy lies close to the top of the barrier, but it is important also for energies well below the top, if one wants to obtain very accurate results with the use of higher orders of the phase-integral approximation. Under the assumption that $d^2Q^2(z)/dz^2$ is not too close to zero at the top of the barrier, Fröman, Fröman and Lundborg (1996) derived for a complex potential barrier by means of comparison equation technique, adapted to yield formulas for supplementary quantities in the phase-integral method, an approximate, but very accurate, formula in the $(2N+1)$th order of the phase-integral approximation for a quantity ϕ [their Eqs. (5.5.30), (5.5.25a–g), (5.4.23) and (5.4.21)], from which one for a real potential barrier can obtain the following formula, as described in Section 3.45 of Fröman and Fröman (2002),

$$\tilde{\phi} = \arg\Gamma\left(\frac{1}{2} + i\bar{K}\right) - \bar{K}\ln|\bar{K}_0| + \sum_{n=0}^{N} \phi^{(2n+1)}, \tag{4.46}$$

where according to Eqs. (5.5.25a–c), (5.4.23) and (5.4.21) with $\lambda = 1$ in Fröman, Fröman and Lundborg (1996)

$$\phi^{(1)} = \bar{K}_0, \tag{4.47a}$$

$$\phi^{(3)} = -\frac{1}{24\bar{K}_0}, \tag{4.47b}$$

$$\phi^{(5)} = -\frac{7}{2880\bar{K}_0^3} + \frac{\bar{K}_2}{24\bar{K}_0^2} - \frac{\bar{K}_2^2}{2\bar{K}_0}, \tag{4.47c}$$

with

$$\bar{K} = \sum_{n=0}^{N} \bar{K}_{2n} = \frac{K}{\pi}, \tag{4.48a}$$

$$\bar{K}_{2n} = \frac{1}{2\pi i}\int_\Lambda Y_{2n}Q(z)\,dz, \quad n = 0, 1, 2, \ldots, N, \tag{4.48b}$$

Λ being a contour of integration encircling t' and t'' but no other transition point, with the integration performed in the direction that makes $\bar{K}_0$ positive when t' and t'' are real, i.e., when the barrier is superdense, but negative when t' and t'' are complex conjugate, i.e., when the barrier is underdense.

We emphasize that for the validity of (4.46) with the expressions (4.47a–c) for $\phi^{(2n+1)}$ the essential restriction is that $|d^2Q^2(z)/dz^2|$ must not be too small at the top of the barrier, which means that close to its top the barrier is approximately parabolic, i.e., that the distance from the barrier to the transition points that are not associated with the barrier must be much larger than $|t'' - t'|$. However, when the energy is close to the top of the barrier, it is the slight deviation from parabolic shape close to the top that determines the values of the quantities $\bar{K}_{2n}, n > 0$, and one needs accurate values of these quantities for obtaining accurate values of $\tilde{\phi}$ in higher orders of the phase-integral approximation.

The derivative of $\tilde{\phi}$ with respect to the energy E will be needed in Chapter 5, and therefore we shall now give formulas for this derivative. From (4.46) and (4.47a–c) one obtains

$$\frac{d\tilde{\phi}}{dE} = \left(\frac{d\arg\Gamma\left(\frac{1}{2}+i\bar{K}\right)}{d\bar{K}} - \ln|\bar{K}_0|\right)\frac{d\bar{K}}{dE} - \frac{\bar{K}}{\bar{K}_0}\frac{d\bar{K}_0}{dE} + \sum_{n=0}^{N}\frac{d\phi^{(2n+1)}}{dE} \tag{4.49}$$

with

$$\frac{d\phi^{(1)}}{dE} = \frac{d\bar{K}_0}{dE}, \tag{4.50a}$$

$$\frac{d\phi^{(3)}}{dE} = \frac{1}{24\bar{K}_0^2}\frac{d\bar{K}_0}{dE}, \tag{4.50b}$$

$$\frac{d\phi^{(5)}}{dE} = \left(\frac{7}{960\bar{K}_0^4} - \frac{\bar{K}_2}{12\bar{K}_0^3} + \frac{\bar{K}_2^2}{2\bar{K}_0^2}\right)\frac{d\bar{K}_0}{dE} + \left(\frac{1}{24\bar{K}_0^2} - \frac{\bar{K}_2}{\bar{K}_0}\right)\frac{d\bar{K}_2}{dE}. \tag{4.50c}$$

According to sections 6.1.27 and 6.3.3 in Abramowitz and Stegun (1965) the argument of the gamma function occurring in (4.49) can be obtained from the formula

$$\arg\Gamma\left(\frac{1}{2}+i\bar{K}\right) = \sum_{n=0}^{\infty}\left(\frac{\bar{K}}{\frac{1}{2}+n} - \arctan\frac{\bar{K}}{\frac{1}{2}+n}\right) - (\gamma+2\ln 2)\bar{K}, \tag{4.51}$$

where $\gamma = \Gamma'(1)/\Gamma(1)$ is Euler's constant. From (4.51) we obtain

$$\frac{d\arg\Gamma\left(\frac{1}{2}+i\bar{K}\right)}{d\bar{K}} = \left\{\bar{K}^2\sum_{n=0}^{\infty}\frac{1}{\left(n+\frac{1}{2}\right)\left[\left(n+\frac{1}{2}\right)^2+\bar{K}^2\right]} - (\gamma+2\ln 2)\right\}\frac{d\bar{K}}{dE}. \tag{4.52}$$

The barrier connection formula presented in this section is valid uniformly for all energies, below and somewhat above the top of the barrier. We would also like to emphasize that while the connection formulas pertaining to a turning point are one-directional (N. Fröman 1966a, Fröman and Fröman 2002), the barrier connection formula (4.30a,b) along with (4.31)–(4.35) is bi-directional. However, when the energy is close to a resonance energy, a careful discussion is required.

4.4 Quantization conditions for single-well potentials

In this section we shall present quantization conditions for general single-well potentials [N. Fröman (1966c, 1978), Fröman and Fröman (1965, 1978a, 1978b, 1978c, 1996, 2002), Paulsson, Karlsson and LeRoy (1983)], valid for any conveniently chosen order of the phase-integral approximation, in forms especially adapted to the treatment of the Stark effect in a hydrogenic atom or ion.

We assume that $R(z)$ and $Q^2(z)$ are real on the real z-axis (the x-axis) and that $R(z)$ is given by (4.23), i.e.,

$$R(z) = -\frac{l(l+1)}{z^2} + \frac{B}{z} + \begin{array}{l}\text{a function of } z \text{ that is regular}\\ \text{in a region around the origin}\end{array}, \tag{4.53}$$

and that

$$Q^2(z) = R(z) - \frac{1}{4z^2}. \tag{4.54}$$

When $l \neq -1/2$ there is on the positive part of the real axis a classically forbidden region in the generalized sense delimited by the origin. It is assumed to be delimited also by a generalized classical turning point t', i.e., a first-order zero of $Q^2(z)$. In this region the wave function is

$$\psi(x) = \text{const} \times |q^{-1/2}(x)| \exp\left[-\left|\int_{(t')}^{x} q(z)\,dz\right|\right], \tag{4.55}$$

and this expression for the wave function remains valid close to the origin. With the use of (4.55) and the connection formula (4.19) one finds that in the classically allowed region (in the generalized sense) to the right of t' the wave function is

$$\psi(x) = \text{const} \times |q^{-1/2}(x)| \cos\left[\left|\int_{(t')}^{x} q(z)\,dz\right| - \frac{\pi}{4}\right]. \tag{4.56}$$

We assume that the classically allowed region delimited by t' is also delimited by another turning point t'' [simple zero of $Q^2(z)$], to the right of which there is a classically forbidden region extending to $+\infty$. In this region the wave function is

$$\psi(x) = \text{const} \times |q^{-1/2}(x)| \exp\left[-\left|\int_{(t'')}^{x} q(z)\,dz\right|\right]. \tag{4.57}$$

By means of (4.57) and the connection formula (4.19) one finds that the physically acceptable wave function is given by

$$\psi(x) = \text{const} \times |q^{-1/2}(x)| \cos\left[\left|\int_{(t'')}^{x} q(z)dz\right| - \pi/4\right] \tag{4.58}$$

in the classically allowed region (in the generalized sense) to the left of t''. By identifying (4.56) and (4.58) one obtains the quantization condition

$$\left|\int_{(t')}^{(t'')} q(z)dz\right| = \left(s + \frac{1}{2}\right)\pi, \quad s = 0, 1, 2, \ldots, \tag{4.59}$$

i.e.,

$$\left|\frac{1}{2}\int_{\Lambda_L} q(z)dz\right| = \left(s + \frac{1}{2}\right)\pi, \quad s = 0, 1, 2, \ldots, \tag{4.60}$$

where Λ_L is a closed contour of integration that encircles both t' and t'' but no other transition points.

When $I = -1/2$ and B is positive, there is on the positive part of the real axis a generalized classically allowed region $[Q^2(x) > 0]$ delimited to the left by the origin. In this region the wave function is according to (4.29) given by the formula

$$\psi(x) = \text{const} \times |q^{-1/2}(x)| \cos\left[\left|\int_{(0)}^{x} q(z)dz\right| - \frac{\pi}{4}\right], \tag{4.61}$$

when B is sufficiently large, and x lies sufficiently far away from the origin. We assume that the classically allowed region delimited by the origin is also delimited by a generalized classical turning point t'' [simple zero of $Q^2(z)$], to the right of which there is a classically forbidden region extending to $+\infty$. In the classically allowed region the wave function is then given by (4.58). Identifying (4.58) and (4.61), we obtain the quantization condition

$$\left|\int_{(0)}^{(t'')} q(z)dz\right| = \left(s + \frac{1}{2}\right)\pi, \quad s = 0, 1, 2, \ldots, \tag{4.62}$$

i.e.,

$$\left|\frac{1}{2}\int_{\Lambda_L} q(z)dz\right| = \left(s + \frac{1}{2}\right)\pi, \quad s = 0, 1, 2, \ldots, \tag{4.63}$$

where Λ_L is a closed contour of integration encircling the origin and t'' but no other transiton points.

With the base function chosen according to (4.54), it is seen from (4.60) and (4.63) that one obtains the same quantization condition whether $l \neq -1/2$ or $l = -1/2$. It should, however, be emphasized that the motivation for the formulas, on which the quantization condition is based, is quite different in the cases when $l \neq -1/2$ and $l = -1/2$.

Chapter 5

Derivation of Phase-Integral Formulas for Profiles, Energies and Half-Widths of Stark Levels

In Chapter 3 we investigated the development in time of a decaying state, expressed in terms of the time-independent eigenfunctions satisfying a system of two coupled differential equations, resulting from the separation of the Schrödinger equation in parabolic coordinates. In this analysis we obtained general expressions for the time-dependent wave function and the probability amplitude.

In the present chapter we shall start from the results obtained in Chapter 3 and treat the Stark effect of a hydrogenic atom or ion with the use of the phase-integral approximation generated from an unspecified base function developed by the present authors and briefly described in Chapter 4 of this book. Phase-integral formulas for profiles, energies and half-widths of Stark levels are obtained. The profile has a Lorentzian shape when the level is narrow but a non-Lorentzian shape when the level is broad. A formula for the half-width is derived on the assumption that the level is not too broad.

For $m \neq 0$ as well as for $m = 0$ it is convenient to choose [cf. (2.31b) and (2.32b)]

$$\tilde{Q}^2(\xi) = \tilde{R}(\xi) - \frac{1}{4\xi^2} = \frac{\mu E}{2\hbar^2} + \frac{Z_1}{\xi} - \frac{m^2}{\xi^2} - \frac{\mu e F \xi}{4\hbar^2}, \tag{5.1}$$

$$Q^2(\eta) = R(\eta) - \frac{1}{4\eta^2} = \frac{\mu E}{2\hbar^2} + \frac{Z_2}{\eta} - \frac{m^2}{\eta^2} + \frac{\mu e F \eta}{4\hbar^2}. \tag{5.2}$$

The roots of $\tilde{Q}^2(\xi)$ are called ξ_0, ξ_1, ξ_2 and the roots of $Q^2(\eta)$ are called η_1, η_2, η_3. The choice of $\tilde{Q}^2(\xi)$ and $Q^2(\eta)$ according to (5.1) and (5.2) is analogous to the replacement of $l(l+1)$ by $(l+1/2)^2$ in the first-order Carlini (JWKB) approximation associated with the radial Schrödinger equation.

The functions $-\tilde{Q}^2(\xi)$ and $-Q^2(\eta)$ are qualitatively depicted in Figs. 5.1a–c and Figs. 5.2a–c for the cases $m \neq 0$ and $m = 0$, respectively. The physically acceptable wave functions with the ξ-variable correspond to discrete energy eigenvalues, while with the η-variable there is associated a continuous energy spectrum, unless one encloses the η-variable in a finite region, as we have done in Chapter 2.

For the $(2N+1)$th-order phase-integral approximation we have according to (4.11) with $\lambda = 1$

$$\tilde{q}(\xi) = \tilde{Q}(\xi) \sum_{n=0}^{N} \tilde{Y}_{2n} \tag{5.3}$$

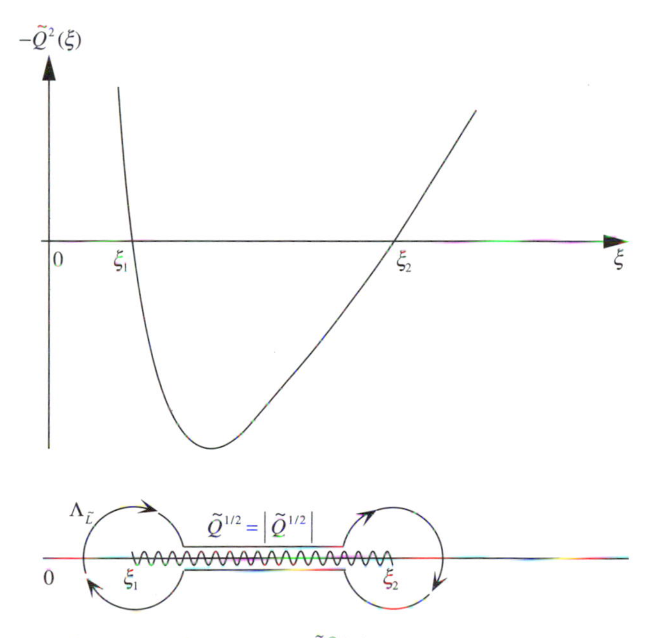

Fig. 5.1a. Qualitative behavior of $-\tilde{Q}^2(\xi)$ for $m \neq 0$. The wavy line indicates a cut, and $\Lambda_{\tilde{L}}$ is a closed contour of integration, on which the phase of $\tilde{Q}^{1/2}(\xi)$ is indicated. The point ξ_0 lies to the left of the origin.

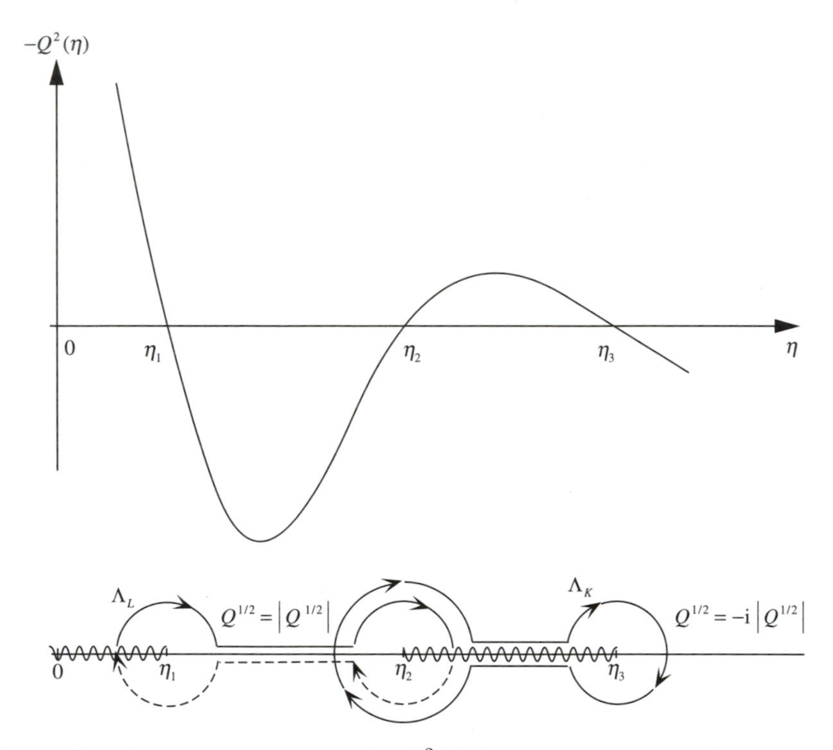

Fig. 5.1b. Qualitative behavior of $-Q^2(\eta)$ for $m \neq 0$ in the sub-barrier case. The wavy lines are cuts, and Λ_L and Λ_K are closed contours of integration. The part of Λ_L that lies on the second Riemann sheet is drawn as a broken line. The phases of $Q^{1/2}(\eta)$ indicated in the figure refer to the first Riemann sheet.

and

$$q(\eta) = Q(\eta) \sum_{n=0}^{N} Y_{2n}, \tag{5.4}$$

where according to (4.10a–c), (4.8) and (4.6)

$$\tilde{Y}_0 = 1, \tag{5.5a}$$

$$\tilde{Y}_2 = \frac{1}{2}\tilde{\varepsilon}_0, \tag{5.5b}$$

$$\tilde{Y}_4 = -\frac{1}{8}\left(\tilde{\varepsilon}_0^2 + \frac{d^2\tilde{\varepsilon}_0}{d\tilde{\zeta}^2}\right), \tag{5.5c}$$

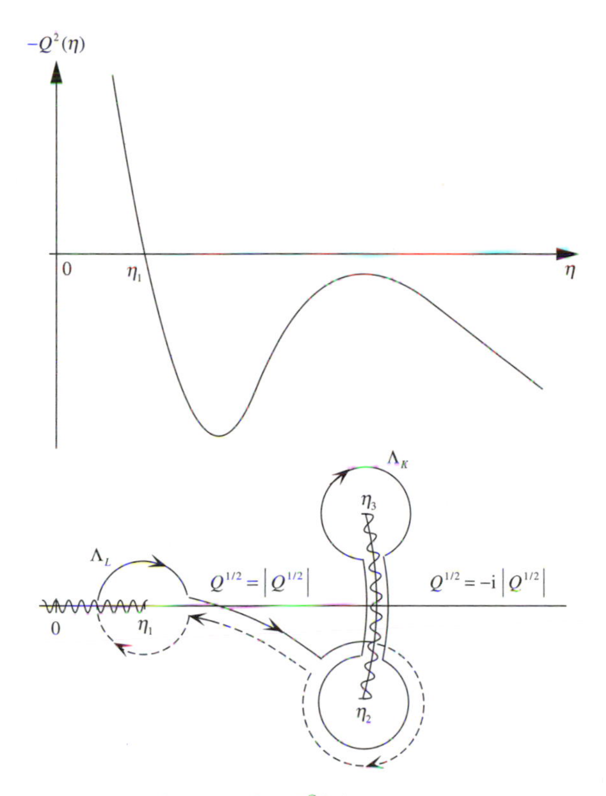

Fig. 5.1c. Qualitative behavior of $-Q^2(\eta)$ for $m \neq 0$ in the super-barrier case. The wavy lines are cuts, and Λ_L and Λ_K are closed contours of integration. The part of Λ_L that lies on the second Riemann sheet is drawn as a broken line. The phases of $Q^{1,2}(\eta)$ indicated in the figure refer to the the first Riemann sheet.

with

$$\tilde{\varepsilon}_0 = \tilde{Q}^{-3/2}(\xi)\frac{d^2}{d\xi^2}\tilde{Q}^{-1/2}(\xi) + \frac{\tilde{R}(\xi) - \tilde{Q}^2(\xi)}{\tilde{Q}^2(\xi)}, \tag{5.6a}$$

$$\tilde{\zeta} = \int^{\xi} \tilde{Q}(\xi)d\xi, \tag{5.6b}$$

and

$$Y_0 = 1, \tag{5.7a}$$

$$Y_2 = \frac{1}{2}\varepsilon_0, \tag{5.7b}$$

$$Y_4 = -\frac{1}{8}\left({\varepsilon_0}^2 + \frac{d^2\varepsilon_0}{d\zeta^2}\right), \tag{5.7c}$$

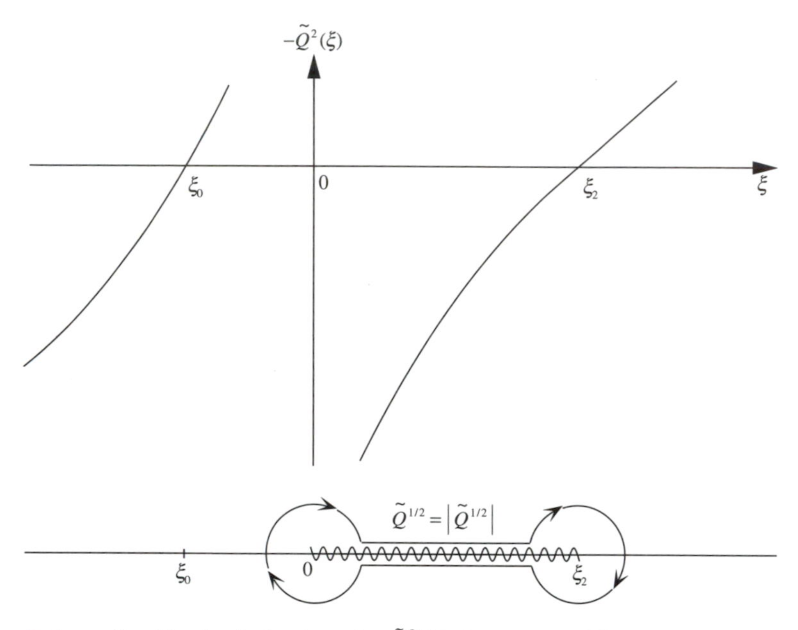

Fig. 5.2a. Qualitative behavior of $-\tilde{Q}^2(\xi)$ for $m = 0$. The wavy line is a cut, and the closed contour of integration, on which the phase of $\tilde{Q}^{1/2}(\xi)$ is indicated, is called $\Lambda_{\tilde{L}}$. The point ξ_1 lies at the origin.

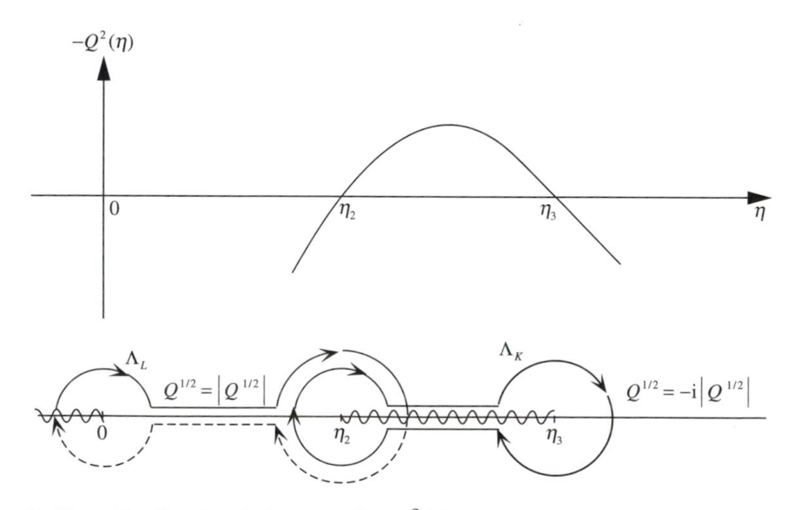

Fig. 5.2b. Qualitative behavior of $-Q^2(\eta)$ for $m = 0$ in the sub-barrier case. The wavy lines are cuts, and Λ_L and Λ_K are closed contours of integration. The part of Λ_L that lies on the second Riemann sheet is drawn as a broken line. The phases of $Q^{1/2}(\eta)$ indicated in the figure refer to the first Riemann sheet. The point η_1 lies at the origin.

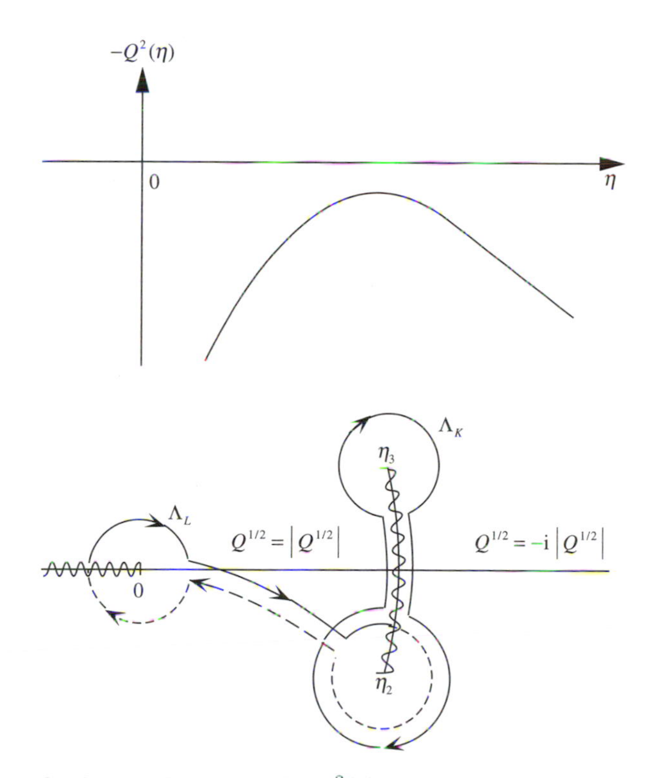

Fig. 5.2c. Qualitative behavior of $-Q^2(\eta)$ for $m = 0$ in the super-barrier case. The wavy lines are cuts, and Λ_L and Λ_K are closed contours of integration. The part of Λ_L that lies on the second Riemann sheet is drawn as a broken line. The phases of $Q^{1/2}(\eta)$ indicated in the figure refer to the first Riemann sheet. The point η_1 lies at the origin.

with

$$\varepsilon_0 = Q^{-3/2}(\eta)\frac{d^2}{d\eta^2}Q^{-1/2}(\eta) + \frac{R(\eta) - Q^2(\eta)}{Q^2(\eta)}, \tag{5.8a}$$

$$\zeta = \int^{\eta} Q(\eta)d\eta. \tag{5.8b}$$

As in the discussion of the time-independent eigenfunctions in Chapter 2, we confine the system such that $0 \leq \xi < \infty$ and $0 \leq \eta \leq \rho$, where ρ is a large quantity which we shall finally let tend to infinity.

With $\tilde{Q}^2(\xi)$ given by (5.1) the quantization condition associated with the differential equation (2.31a,b) is, with the aid of (4.60) when $m \neq 0$ and with the aid of (4.63) when $m = 0$, and with due regard to Figs. 5.1a and 5.2a, found to be

$$\tilde{L} = \left(n_1 + \frac{1}{2}\right)\pi, \tag{5.9}$$

where n_1 is a non-negative integer and [cf. (5.3)]

$$\tilde{L} = \sum_{n=0}^{N} \tilde{L}_{2n}, \tag{5.10a}$$

$$\tilde{L}_{2n} = \int_{\Lambda_L} \tilde{Y}_{2n}\tilde{Q}(\xi)d\xi, \tag{5.10b}$$

with the contour of integration $\Lambda_{\tilde{L}}$ and the phase of $\tilde{Q}(\xi)$ shown in Figs. 5.1a and 5.2a. The justification of the quantization condition (5.9) is quite different in the cases when $m \neq 0$ and when $m = 0$, since (4.60) is derived with the aid of the connection formula (4.19), while (4.63) is derived with the aid of not only the connection formula (4.19) but also the particular case (4.29) of the connection formula (4.28). We also recall the formula (2.46), which determines the normalization of the function $f(m, n_1, E_{m,n_1,s}; \xi)$, i.e.,

$$\int_0^{\infty} [f(m, n_1, E_{m,n_1,s}; \xi)]^2 \frac{d\xi}{4\xi} = 1. \tag{5.11}$$

The only knowledge that is needed about the function $f(m, n_1, E_{m,n_1,s}; \xi)$ is (5.11) and the fact that the function in question is almost independent of ρ for large values of ρ.

With $Q^2(\eta)$ given by (5.2) we recall (4.19) and (4.29) and normalize the physically acceptable solution $g(m, n_1, E_{m,n_1,s}; \eta)$ of the differential equation (2.32a,b) such that in the classically allowed region to the left of the barrier in Figs. 5.1b,c and Figs. 5.2b,c the phase-integral expression for this solution, with the use of the short-hand notation defined in (4.17), is

$$g(m, n_1, E_{m,n_1,s}; \eta) = q^{-1/2}(\eta) \cos\left[\int_{(\eta_1)}^{\eta} q(\eta)d\eta - \frac{\pi}{4}\right], \quad m \neq 0, \tag{5.12a}$$

$$g(m, n_1, E_{m,n_1,s}; \eta) = q^{-1/2}(\eta) \cos\left[\int_{(0)}^{\eta} q(\eta)d\eta - \frac{\pi}{4}\right], \quad m = 0, \tag{5.12b}$$

with $q(\eta)$ given by (5.4) and (5.2) with $E = E_{m,n_1,s}$ and the phase of $Q^{1/2}(\eta)$ chosen as shown in Figs. 5.1b,c and Figs. 5.2b,c. We remark that the expression for the eigenfunction in the well to the left of the barrier, i.e., (5.12a) or (5.12b), is for $m \neq 0$ the result of using the connection formula (4.19), but for $m = 0$ the result of using the particular case (4.29) of the connection formula (4.28). Although the right-hand sides look identical, except for the normalization factor, the justifications for these two formulas are quite different; see Section 4.2. The normalization factor $\Omega(m, n_1, E_{m,n_1,s})$ in (2.33) will be determined later such that the condition (2.47) is fulfilled. To the left of the barrier (5.12a) and (5.12b) can for real values of η be rewritten as

$$\begin{aligned} g(m, n_1, E_{m,n_1,s}; \eta) &= q^{-1/2}(\eta) \cos\left[\mathrm{Re}\int_{\eta}^{(\eta_2)} q(\eta)d\eta + \delta' - \frac{\pi}{4}\right] \\ &= |q^{-1/2}(\eta)| \cos\left[\left|\mathrm{Re}\int_{(\eta_2)}^{\eta} q(\eta)d\eta\right| + \delta' - \frac{\pi}{4}\right], \end{aligned} \tag{5.13}$$

where

$$\delta' = \frac{\pi}{2} - L \tag{5.14}$$

with L defined as [cf. (5.4)]

$$L = \mathrm{Re}\frac{1}{2}\int_{\Lambda_L} q(\eta)d\eta = \sum_{n=0}^{N} L_{2n}, \tag{5.15a}$$

$$L_{2n} = \mathrm{Re}\frac{1}{2}\int_{\Lambda_L} Y_{2n}Q(\eta)d\eta, \tag{5.15b}$$

Λ_L being the contour of integration shown in Figs. 5.1b,c when $m \neq 0$ and in Figs. 5.2b,c when $m = 0$. According to the connection formula (4.38a,b) for a real potential barrier the particular solution of the differential equation (2.32a,b), which to the left of the barrier is given by (5.13), is to the right of the barrier [cf. Figs. 5.1b,c and Figs. 5.2b,c] given by

$$g(m, n_1, E_{m,n_1,s}; \eta) = \frac{\Omega''}{\Omega'} |q^{-1/2}(\eta)| \cos \left[\left| \mathrm{Re} \int_{(\eta_3)}^{\eta} q(\eta) d\eta \right| + \delta'' - \frac{\pi}{4} \right] \tag{5.16}$$

with δ'' and Ω'' given by (4.43a,b), i.e.,

$$\delta'' = \arctan \left[\frac{u-1}{u+1} \tan \left(\frac{\pi}{2} - v \right) \right] + \frac{\tilde{\phi}}{2} - \frac{\vartheta}{2}, \tag{5.17a}$$

$$\begin{aligned} \Omega'' &= \Omega' \left[\frac{u-1}{u+1} + \frac{4u}{u^2-1} \sin^2 v \right]^{1/2} \\ &= \Omega' \left[\frac{u+1}{u-1} - \frac{4u}{(u^2-1)(1+\tan^2 v)} \right]^{1/2}, \end{aligned} \tag{5.17b}$$

where according to (4.44a,b), (4.33), (4.35) and (5.14)

$$u = (1 + 1/\theta^2)^{1/2} \approx [1 + \exp(-2K)]^{1/2}, \tag{5.18a}$$

$$v = \delta' - \frac{\tilde{\phi}}{2} - \frac{\vartheta}{2} = \frac{\pi}{2} - L - \frac{\tilde{\phi}}{2} - \frac{\vartheta}{2} \approx \frac{\pi}{2} - L - \frac{\tilde{\phi}}{2}. \tag{5.18b}$$

The quantity L is given by (5.15a,b), and the quantity K is given by (4.48a,b), i.e.,

$$K = \pi \bar{K} = \pi \sum_{n=0}^{N} \bar{K}_{2n}, \tag{5.19a}$$

$$\bar{K}_{2n} = \frac{i}{2\pi} \int Y_{2n} Q(\eta) d\eta, \tag{5.19b}$$

the contour of integration Λ_K being depicted in Figs. 5.1b,c [$m \neq 0$] and Figs. 5.2b,c [$m = 0$], and the integration along it being performed in such a direction that $\bar{K}_0$ is positive when the barrier is superdense but negative when the barrier is underdense. We remark that (4.48a,b) and (5.19a,b) are consistent, since the directions of integration along Λ and Λ_K are different.

For large values of $\eta(\leq \rho)$ the cosine in (5.16) is a strongly oscillating function of η, and with the use of (5.2) we therefore obtain

$$\begin{aligned}
&\int_0^\rho [g(m, n_1, E_{m,n_1,s}; \eta)]^2 d\eta \\
&\quad \approx \int_{(\eta_3)}^\rho \left(\frac{\Omega''}{\Omega'}\right)^2 \cos^2\left[\left|\mathrm{Re}\int_{(\eta_3)}^\eta q(\eta)d\eta\right| + \delta'' - \frac{\pi}{2}\right]\frac{d\eta}{|q(\eta)|} \\
&\quad = \left(\frac{\Omega''}{\Omega'}\right)^2 \int_{(\eta_3)}^\rho \left\{1 + \cos\left[2\left(\left|\mathrm{Re}\int_{(\eta_3)}^\eta q(\eta)d\eta\right| + \delta'' - \frac{\pi}{4}\right)\right]\right\}\frac{d\eta}{2|q(\eta)|} \\
&\quad \approx \left(\frac{\Omega''}{\Omega'}\right)^2 \int_{(\eta_3)}^\rho \frac{d\eta}{2|q(\eta)|} \approx \left(\frac{\Omega''}{\Omega'}\right)^2 \int_{(\eta_3)}^\rho \frac{1}{2}\left(\frac{\mu e F\eta}{4\hbar^2}\right)^{-1/2} d\eta \\
&\quad = \left(\frac{\Omega''}{\Omega'}\right)^2 \left(\frac{4\hbar^2}{\mu e F}\right)^{1/2} \int_{(\eta_3)}^\rho \frac{d\eta}{2\eta^{1/2}} \approx \left(\frac{\Omega''}{\Omega'}\right)^2 \left(\frac{4\hbar^2 \rho}{\mu e F}\right)^{1/2},
\end{aligned}$$

i.e.,

$$\int_0^\rho [g(m, n_1, E_{m,n_1,s}; \eta)]^2 d\eta \approx \left(\frac{\Omega''}{\Omega'}\right)^2 \left(\frac{4\hbar^2 \rho}{\mu e F}\right)^{1/2}. \tag{5.20}$$

In a similar way one finds that $\int_0^\rho [g(m, n_1, E_{m,n_1,s}; \eta)]^2 d\eta/\eta$ is equal to $(\Omega''/\Omega')^2$ times a factor that is approximately independent of ρ in the limit when $\rho \to \infty$. This, together with (5.20), justifies in another way the approximation of (2.36) that leads to (2.45), which is valid when ρ is sufficiently large. Inserting (5.20) into (2.47), we obtain

$$[\Omega(m, n_1, E_{m,n_1,s})]^2 \approx \left(\frac{\mu e F}{4\hbar^2 \rho}\right)^{1/2} \left(\frac{\Omega'}{\Omega''}\right)^2. \tag{5.21}$$

The normalization factor $\Omega(m, n_1, E_{m,n_1,s})$ thus depends on ρ, while $g(m, n_1, E_{m,n_1,s}; \eta)$ does not depend on ρ; see (5.12a,b).

The condition that $g(m, n_1, E_{m,n_1,s}; \eta)$ be equal to zero when η is equal to the very large quantity ρ implies according to (5.16) that

$$\left|\mathrm{Re}\int_{(\eta_3)}^\rho q(\eta)d\eta\right| + \delta'' - \frac{\pi}{4} = \frac{\pi}{2} + \text{an integer multiple of } \pi. \tag{5.22}$$

For large values of $\eta(\leq\rho)$ one obtains from (5.4), (5.5a) and (5.2)

$$q(\eta) \approx Q(\eta) = \left[\frac{\mu eF\eta}{4\hbar^2} + \frac{\mu E}{2\hbar^2} + O(\eta^{-1})\right]^{1/2} \approx \left(\frac{\mu eF\eta}{4\hbar^2}\right)^{1/2} + \left(\frac{\mu}{4\hbar^2 eF\eta}\right)^{1/2} E + O(\eta^{-3/2}), \quad \eta \text{ large}, \tag{5.23}$$

and hence

$$\int_{(\eta_3)}^{\rho} q(\eta)d\eta \approx \left(\frac{\mu eF\rho^3}{9\hbar^2}\right)^{1/2} + \left(\frac{\mu\rho}{\hbar^2 eF}\right)^{1/2} E. \tag{5.24}$$

From (5.22) and (5.23) it follows that the spacing $\Delta E = E_{m,n_1,s+1} - E_{m,n_1,s}$ between two neighboring energy levels $E_{m,n_1,s+1}$ and $E_{m,n_1,s}$ is approximately

$$\Delta E \approx \pi \left(\frac{\hbar^2 eF}{\mu\rho}\right)^{1/2}. \tag{5.25}$$

From (5.21) and (5.25) we obtain

$$\frac{[\Omega(m, n_1, E_{m,n_1,s})]^2}{\Delta E} \approx \frac{\mu}{2\pi\hbar^2}\left(\frac{\Omega'}{\Omega''}\right)^2. \tag{5.26}$$

The quotient $[\Omega(m, n_1, E_{m,n_1,s})]^2/\Delta E$ is thus independent of ρ, since according to (5.17b) and (5.18a,b) Ω'/Ω'' is independent of ρ.

Using (5.26), we can write (3.12) as

$$\bar{C}(m, n_1, E) = \frac{\mu}{2\pi\hbar^2}\left(\frac{\Omega'}{\Omega''}\right)^2 \iiint \bar{\chi}^*(m, n_1, E; x, y, z) \times \psi(x, y, z; 0)\,dx\,dy\,dz, \tag{5.27}$$

(3.13) as

$$\sum_{m=-\infty}^{\infty}\sum_{n_1=0}^{\infty} \frac{2\pi\hbar^2}{\mu}\int \left(\frac{\Omega''}{\Omega'}\right)^2 |\bar{C}(m, n_1, E)|^2 dE = 1, \tag{5.28}$$

and (3.14) as

$$p(t) = \sum_{m=-\infty}^{\infty}\sum_{n_1=0}^{\infty} \frac{2\pi\hbar^2}{\mu}\int \left(\frac{\Omega''}{\Omega'}\right)^2 |\bar{C}(m, n_1, E)|^2 \exp\left(\frac{-iEt}{\hbar}\right) dE. \tag{5.29}$$

By means of (5.28) and (5.29) one can easily confirm that $p(0) = 1$. Inserting (5.27) into (5.29), we obtain

$$p(t) = \sum_{m=-\infty}^{\infty} \sum_{n_1=0}^{\infty} \frac{\mu}{2\pi\hbar^2} \times \int \left(\frac{\Omega'}{\Omega''}\right)^2 \left| \iiint \bar{\chi}^*(m, n_1, E; x, y, z) \right.$$
$$\left. \times \psi(x, y, z; 0) dx\, dy\, dz \right|^2 \exp\left(-\frac{iEt}{\hbar}\right) dE, \tag{5.30}$$

where according to (2.34) with $E_{m,n_1,s}$ replaced by E

$$\bar{\chi}(m, n_1, E; x, y, z) = \frac{f(m, n_1, E; \xi)}{\xi^{1/2}} \frac{g(m, n_1, E; \eta)}{\eta^{1/2}} \frac{\exp(im\varphi)}{(2\pi)^{1/2}} \tag{5.31}$$

with $f(m, n_1, E; \xi)$ normalized according to (5.11) with $E_{m,n_1,s}$ replaced by E and with $g(m, n_1, E; \eta)$ in the classically allowed region to the left of the barrier given by (5.12a,b) with $E_{m,n_1,s}$ replaced by E. According to (5.17b) and (5.18a,b) the energy dependence of the integral in (5.30) is negligible compared to the energy dependence of $(\Omega'/\Omega'')^2$ when K is sufficiently large, and then $(\Omega'/\Omega'')^2$ in (5.30) determines the profile of the Stark levels. The shape and half-width of such a level is thus determined by the profile $(\Omega'/\Omega'')^2$ as function of E, and the position of the level is naturally defined as the energy for which the profile assumes its maximum value. This is the case also when K is not sufficiently large, if one defines the profile not by means of the Fock–Krylov theorem but by means of the function $g(m, n_1, E_{m,n_1,s}; \eta)$ in (5.16).

From (5.17a,b) and (5.18a,b) it is seen that if $\exp(2K) \gg 1$, and if the energy E increases continuously, which according to (5.14) and (5.15a,b) means that δ' decreases continuously, the quantity $(\Omega'/\Omega'')^2$ passes through sharp maxima at which v is equal to an integer multiple of π, and δ'' increases steeply by π, when E passes through such a maximum. When the energy increases, the quantity $\exp(2K)$ decreases, and when E approaches and passes through the top of the barrier, the profile $(\Omega'/\Omega'')^2$ becomes broader and gradually fades out for energies above the top of the barrier. At the same time the profile loses its original Lorentzian shape. The broad levels located close to or above the top of the barrier are highly asymmetric. For a precise characterization of those levels the full profile, calculated from (5.17b) along with (5.18a,b), must be used.

5.1 Positions of the Stark levels

As already mentioned, we define the positions of the Stark levels as the energies for which $(\Omega'/\Omega'')^2$ assumes its maxima for fixed F, m and n_1. When the energy dependence of the u-dependent quantities in (5.17b) is much smaller than that of $\tan v$, it is seen that the resonances, i.e., the minima of $(\Omega''/\Omega')^2$, occur when approximately

$$\tan v = 0. \tag{5.32}$$

According to the approximate version of (5.18b) it follows from (5.32) that

$$L = \left(n_2 + \frac{1}{2}\right)\pi - \frac{\tilde{\phi}}{2}, \tag{5.33}$$

where n_2 is an integer. To obtain a more accurate formula for the positions of the Stark levels we shall now calculate the energy derivative of $(\Omega''/\Omega')^2$ when the effective electric field strength F and the quantum numbers m and n_1 are kept fixed. We obtain from (5.17b)

$$\frac{d}{dE}\left(\frac{\Omega''}{\Omega'}\right)^2 = -\frac{2du/dE}{(u-1)^2(1+\tan^2 v)} \times \left[\tan^2 v - \frac{4u(u-1)dv/dE}{(u+1)du/dE}\tan v - \left(\frac{u-1}{u+1}\right)^2\right], \tag{5.34}$$

from the approximate version of (5.18a)

$$\frac{du}{dE} \approx -\frac{\exp(-2K)dK/dE}{[1+\exp(-2K)]^{1/2}} \approx -\frac{(u+1)(u-1)dK/dE}{u} \tag{5.35}$$

and from the approximate version of (5.18b)

$$\frac{dv}{dE} \approx -\frac{1}{2}\frac{d(2L+\tilde{\phi})}{dE}. \tag{5.36}$$

Inserting (5.35) and (5.36) into (5.34), we get

$$\frac{d}{dE}\left(\frac{\Omega''}{\Omega'}\right)^2 = \frac{2(u+1)dK/dE}{u(u-1)(1+\tan^2 v)} \times \left[\tan^2 v - \frac{2u^2 d(2L+\tilde{\phi})/dE}{(u+1)^2 dK/dE}\tan v - \left(\frac{u-1}{u+1}\right)^2\right]. \tag{5.37}$$

The values of v for which $(\Omega''/\Omega)^2$ assumes its maxima and minima are obtained from the equation

$$\tan^2 v - \frac{2u^2 d(2L+\tilde{\phi})/dE}{(u+1)^2 dK/dE} \tan v - \left(\frac{u-1}{u+1}\right)^2 = 0, \tag{5.38}$$

the solutions of which are

$$\tan v = \frac{u^2 d(2L+\tilde{\phi})/dE}{(u+1)^2 dK/dE} \left\{ 1 \pm \left[1 + \left(\frac{(u^2-1)dK/dE}{u^2 d(2L+\tilde{\phi})/dE} \right)^2 \right]^{1/2} \right\}, \tag{5.39}$$

where the upper and lower signs correspond to maxima and minima, respectively, of $(\Omega''/\Omega')^2$, which one realizes by noting that (5.32) is obtained as an approximation of (5.39) when $K \gg 1$ and one chooses the minus sign in (5.39). From (5.39) one thus finds that the resonances are obtained from the formula

$$\tan v = \Delta, \tag{5.40}$$

where

$$\begin{aligned} \Delta &= \frac{u^2 d(2L+\tilde{\phi})/dE}{(u+1)^2 dK/dE} \left\{ 1 - \left[1 + \left(\frac{(u^2-1)dK/dE}{u^2 d(2L+\tilde{\phi})/dE} \right)^2 \right]^{1/2} \right\} \\ &= -\frac{(u^2-1)^2 dK/dE}{u^2(u+1)^2 d(2L+\tilde{\phi})/dE} \\ &\quad \times \left\{ 1 + \left[1 + \left(\frac{(u^2-1)dK/dE}{u^2 d(2L+\tilde{\phi})/dE} \right)^2 \right]^{1/2} \right\}^{-1}. \end{aligned} \tag{5.41}$$

With the use of the approximate version of (5.18a) we can write (5.41) as

$$\begin{aligned} \Delta = &-\frac{dK/dE}{[\exp(2K)+1]\{[\exp(2K)+1]^{1/2}+\exp(K)\}^2 d(2L+\tilde{\phi})/dE} \\ &\times \left\{ 1 + \left[1 + \left(\frac{dK/dE}{[\exp(2K)+1] d(2L+\tilde{\phi})/dE} \right)^2 \right]^{1/2} \right\}^{-1}. \end{aligned} \tag{5.42}$$

When the barrier is thick, we can write (5.42) approximately as

$$\Delta \approx -\frac{\exp(-4K)dK/dE}{8d(2L+\tilde{\phi})/dE}, \quad \text{thick barrier.} \tag{5.43}$$

With the use of the approximate version of (5.18b) we obtain from (5.40) the following improvement of (5.33)

$$L = \left(n_2 + \frac{1}{2}\right)\pi - \frac{\tilde{\phi}}{2} - \arctan\Delta, \tag{5.44}$$

where n_2 is an integer.

The positions of the Stark levels E_n, where $n = |m| + 1 + n_1 + n_2$, are obtained from the two simultaneous quantization conditions (5.9) and (5.44) along with (5.42).

5.2 Formulas for the calculation of dL/dE, dK_{2n}/dE and dK/dE

The phase-integrals $\tilde{L}$, L and K have besides the explicit energy dependence an implicit energy dependence, since Z_1 and Z_2 change when the energy changes. When the effective electric field strength F and the quantum numbers m and n_1 are kept fixed, while E changes, we obtain from the quantization condition (5.9) and the relation (2.29), according to which $Z_1 + Z_2$ is equal to a constant,

$$\begin{aligned} 0 = \frac{d\tilde{L}(E,Z_1)}{dE} &= \frac{\partial\tilde{L}(E,Z_1)}{\partial E} + \frac{\partial\tilde{L}(E,Z_1)}{\partial Z_1}\frac{dZ_1}{dE} \\ &= \frac{\partial\tilde{L}(E,Z_1)}{\partial E} - \frac{\partial\tilde{L}(E,Z_1)}{\partial Z_1}\frac{dZ_2}{dE} \end{aligned} \tag{5.45}$$

and hence

$$\frac{dZ_2}{dE} = \frac{\partial\tilde{L}(E,Z_1)/\partial E}{\partial\tilde{L}(E,Z_1)/\partial Z_1}, \quad F, m \text{ and } n_1 \text{ fixed.} \tag{5.46}$$

It follows from the first-order approximation of (5.10a,b), (5.5a) and (5.1) that $\partial\tilde{L}(E,Z_1)/\partial E > 0$ and $\partial\tilde{L}(E,Z_1)/\partial Z_1 > 0$, then from (5.46) that $dZ_2/dE > 0$, and finally from (2.29) that $dZ_1/dE < 0$. When F, m and n_1 are kept fixed, we obtain with the use of (5.46) [cf. (5.15a,b)]

$$\begin{aligned} \frac{dL_{2n}(E,Z_2)}{dE} &= \frac{\partial L_{2n}(E,Z_2)}{\partial E} + \frac{\partial L_{2n}(E,Z_2)}{\partial Z_2}\frac{dZ_2}{dE} = \frac{\partial L_{2n}(E,Z_2)}{\partial E} \\ &\quad + \frac{\partial L_{2n}(E,Z_2)}{\partial Z_2}\frac{\partial\tilde{L}(E,Z_1)/\partial E}{\partial\tilde{L}(E,Z_1)/\partial Z_1}. \end{aligned} \tag{5.47}$$

Similarly we obtain

$$\frac{d\bar{K}_{2n}(E,Z_2)}{dE} = \frac{\partial \bar{K}_{2n}(E,Z_2)}{\partial E} + \frac{\partial \bar{K}_{2n}(E,Z_2)}{\partial Z_2} \frac{\partial \tilde{L}(E,Z_1)/\partial E}{\partial \tilde{L}(E,Z_1)/\partial Z_1}. \tag{5.48}$$

We then obtain with the use of (5.15a) and (5.47)

$$\frac{dL}{dE} = \sum_{n=0}^{N} \frac{dL_{2n}(E,Z_2)}{dE} = \frac{\partial L(E,Z_2)}{\partial E} + \frac{\partial L(E,Z_2)}{\partial Z_2} \frac{\partial \tilde{L}(E,Z_1)/\partial E}{\partial \tilde{L}(E,Z_1)/\partial Z_1} \tag{5.49}$$

and with the use of (5.19a) and (5.48)

$$\begin{aligned} \frac{dK}{dE} = \pi \sum_{n=0}^{N} \frac{d\bar{K}_{2n}(E,Z_2)}{dE} &= \frac{\partial K(E,Z_2)}{\partial E} \\ &+ \frac{\partial K(E,Z_2)}{\partial Z_2} \frac{\partial \tilde{L}(E,Z_1)/\partial E}{\partial \tilde{L}(E,Z_1)/\partial Z_1}. \end{aligned} \tag{5.50}$$

By means of (4.49), (4.50a–c), (4.52), (5.48), (5.49) and (5.50) one can calculate the derivatives in (5.42) and (5.43).

5.3 Half-widths of the Stark levels

We shall next give an explicit formula for the half-width Γ on the energy scale of a not too broad Stark level. To this purpose we write (5.17b) with the use of the approximate version of (5.18a) as

$$\begin{aligned} \left(\frac{\Omega'}{\Omega''}\right)^2 &= \frac{\dfrac{u+1}{u-1}}{1 + \dfrac{4u}{(u-1)^2} \sin^2 v} \\ &= \frac{\dfrac{(u+1)^2}{u^2-1}}{1 + \dfrac{\sin^2 v}{(u^{1/2}/2 - u^{-1/2}/2)}} \\ &\approx \frac{\exp(2K)\{[1+\exp(-2K)]^{1/2}+1\}^2}{1 + \dfrac{\sin^2 v}{\left\{\dfrac{1}{2}[1+\exp(-2K)]^{1/4} - \dfrac{1}{2}[1+\exp(-2K)]^{1/4}\right\}^2}}. \end{aligned} \tag{5.51}$$

For a not too broad Stark level an adequate approximate formula for Γ is obtained when one neglects the change with energy of u over the width of the level. Thus one finds from (5.51) that $(\Omega'/\Omega'')^2$ assumes half of its maximum value when

$$|\sin v| \approx \frac{1}{2}[1+\exp(-2K)]^{1/4} - \frac{1}{2}[1+\exp(-2K)]^{1/4}. \tag{5.52}$$

The half-width expressed in terms of the variable v is thus $2\arcsin\{[1+\exp(-2K)]^{1/4}/2 - [1+\exp(-2K)]^{-1/4}/2\}$. By multiplying this quantity by $|dE/dv|$, i.e., by $|dv/dE|^{-1}$, which according to the approximate version of (5.18b) is equal to $|d(L+\tilde{\phi}/2)/dE|^{-1}$, we obtain the half-width Γ, which is thus

$$\Gamma = \frac{4\arcsin\left\{\frac{1}{2}[1+\exp(-2K)]^{1/4} - \frac{1}{2}[1+\exp(-2K)]^{-1/4}\right\}}{|d(2L+\tilde{\phi})/dE|_{E=E_n}}. \tag{5.53}$$

The quantity $d\tilde{\phi}/dE$ in (5.53) is important when the energy approaches the top of the barrier, since it there cancels the singularity in $d(2L)/dE$. We can write (5.53) as

$$\Gamma = \frac{4\arcsin\left\{\frac{1}{2}[1+\exp(-2K)]^{1/4} - \frac{1}{2}[1+\exp(-2K)]^{-1/4}\right\}}{|\tau/\hbar + d\tilde{\phi}/dE|_{E=E_n}}, \tag{5.54}$$

where

$$\tau = 2\hbar\frac{dL}{dE} \tag{5.55}$$

is the time for a complete classical oscillation to and fro in the well to the left of the barrier. In the first-order approximation $\tilde{\phi} \approx -\phi^{(3)} = 1/(24\bar{K}_0)$ for a thick barrier according to Eqs. (2.5.14) and (2.5.13b) in Fröman and Fröman (2002), and in (5.53) and (5.54) $d\tilde{\phi}/dE$ can therefore be neglected when $d(1/\bar{K}_0)/dE \ll 48dL/dE$.

Chapter 6

Procedure for Transformation of the Phase-Integral Formulas into Formulas Involving Complete Elliptic Integrals

The phase-integral quantities in the formulas obtained in Chapter 5 can be expressed in terms of complete elliptic integrals. One thereby achieves the result that well-known properties of complete elliptic integrals, such as for instance series expansions, can be exploited for analytic studies. Furthermore, complete elliptic integrals can be evaluated very rapidly by means of standard computer programs.

In this chapter we shall describe the procedure for expressing the phase-integral formulas derived in Chapter 5 in terms of complete elliptic integrals. The integral in question is first expressed in terms of a Jacobian elliptic function and then in terms of complete elliptic integrals. Different elliptic functions are appropriate for different phase-integrals. For practical calculations it is most convenient to work with real quantities. For the phase-integrals associated with the η-equation it is therefore appropriate to use different formulas for the sub-barrier case and for the super-barrier case. We indicate in this chapter, where we use the notations $\tilde{L}^{(2n+1)}$, $\tilde{L}^{(2n+1)}$, $K^{(2n+1)}$ instead of the notations $\tilde{L}_{2n}$, L_{2n}, K_{2n} used previously, the main steps in the procedure for expressing, for $m \neq 0$, $L^{(1)}$ both in the sub-barrier case and in the super-barrier case and $K^{(1)}$ in the super-barrier case in terms of complete elliptic integrals.

From now on we shall use units such that $\mu = e = \hbar = 1$, i.e., atomic units when the nucleus is assumed to be infinitely heavy; see (2.5). Denoting the zeros of $Q^2(\eta)$ by η_1, η_2 and η_3, we write (5.2) with $\mu = e = \hbar = 1$ as

$$Q^2(\eta) = \frac{F(\eta - \eta_1)(\eta - \eta_2)(\eta - \eta_3)}{4\eta^2}, \tag{6.1}$$

where $\eta_1 = 0$ when $m = 0$; see Figs. 5.2b and 5.2c.

Considering first $L^{(1)}$ in the sub-barrier case in Fig. 5.1b, we introduce the Jacobian elliptic function snu by the transformation

$$sn^2 u = \frac{\eta - \eta_1}{\eta_2 - \eta_1}, \tag{6.2}$$

i.e.,

$$\eta = \eta_1 + (\eta_2 - \eta_1) sn^2 u. \tag{6.3}$$

Recalling (5.15a,b) and (5.7a), and using (6.1), (6.3) and general properties of Jacobian elliptic functions, we obtain after some calculations

$$\begin{aligned} L^{(1)} = \int_{\eta_1}^{\eta_2} Q(\eta) d\eta &= \frac{F^{1/2}(\eta_2 - \eta_1)^2 (\eta_3 - \eta_1)^{1/2}}{\eta_1} \\ &\times \int_0^K \frac{sn^2 u (1 - sn^2 u)(1 - k^2 sn^2 u)}{1 - \alpha^2 sn^2 u} du, \end{aligned} \tag{6.4}$$

where

$$k^2 = \frac{(\eta_2 - \eta_1)}{(\eta_3 - \eta_1)}, \quad \alpha^2 = -\frac{(\eta_2 - \eta_1)}{\eta_1}. \tag{6.5a,b}$$

After having performed the integration in (6.4), we obtain

$$\begin{aligned} L^{(1)} = -\frac{F^{1/2}(\eta_3 - \eta_1)^{3/2}}{3} k^2 (1 - k^2) \Bigg[\frac{K(k) - E(k)}{k^2} \\ + \left(\frac{3}{\alpha^2} - 2 \right) \frac{E(k)}{1 - k^2} - \frac{3}{\alpha^2} \Pi \left(\frac{k^2 - \alpha^2}{1 - \alpha^2}, k \right) \Bigg], \end{aligned} \tag{6.6}$$

K, E and Π being complete elliptic integrals of the first, second and third kind, respectively. According to Section 117.03 on Page 14 in Byrd and Friedman (1971) we have the formula

$$\Pi\left(\frac{k^2 - \alpha^2}{1 - \alpha^2}, k \right) = \frac{1 - \alpha^2}{\alpha^2 (1 - k^2)} [k^2 K(k) - (k^2 - \alpha^2) \Pi(\alpha^2, k)], \tag{6.7}$$

by means of which (6.6) can be written as

$$L^{(1)} = \frac{F^{1/2}(\eta_3 - \eta_1)^{3/2}}{3}\left\{k^2(1-k^2)\left[\frac{3\left(\frac{k^4}{\alpha^2}\right)\left(\frac{1}{\alpha^2}-1\right)}{1-k^2} - 1\right]\right.$$
$$\times \frac{K(k)-E(k)}{k^2} + k^2(1-k^2)\left[3\left(1-\frac{1}{\alpha^2}\right)\left(1-\frac{k^2}{\alpha^2}\right)-1\right]$$
$$\left.\times \frac{E(k)}{1-k^2} - 3k^2\left(1-\frac{1}{\alpha^2}\right)\left(1-\frac{k^2}{\alpha^2}\right)\Pi(\alpha^2,k)\right\}. \tag{6.8}$$

With the aid of the expressions (7.13c) and (7.13a$'$,b$'$,c$'$) in Chapter 7 for β, k'^2, α'^2 and β' we can write (6.8) as

$$L^{(1)} = \frac{F^{1/2}(\eta_3 - \eta_1)^{3/2}}{3}\left\{k'^2(1-k'^2)[3(1-\beta')(1-k'^2\beta')-1]\right.$$
$$\times \frac{K(k)-E(k)}{k^2} + k^2(1-k^2)[3(1-\beta)(1-k^2\beta)-1]$$
$$\left.\times \frac{E(k)}{k'^2} - 3k^2(1-\beta)(1-k^2\beta)\Pi(\alpha^2,k)\right\}. \tag{6.9}$$

Using (7.3a), (7.6a) and (7.13d) in Chapter 7, we can write (6.9) in the form of (7.14a), with $m \neq 0$, in Chapter 7.

Next we consider $L^{(1)}$ in the super-barrier case in Fig. 5.1c, where the zeros η_2 and η_3 of $Q^2(\eta)$ are complex conjugate. Denoting their real and imaginary parts by b_1 and $\mp a_1$, respectively, we have

$$\eta_2 = b_1 - ia_1, \quad \eta_3 = b_1 + ia_1, \tag{6.10a,b}$$

a_1 being positive according to Fig. 5.1c. We introduce the Jacobian elliptic function cnu by the transformation [see Page 86 in Byrd and Friedman (1971)]

$$cnu = \frac{A + \eta_1 - \eta}{A - \eta_1 + \eta}, \tag{6.11}$$

where

$$A = [(b_1 - \eta_1)^2 + a_1{}^2]^{1/2}. \tag{6.12}$$

From (6.11) we obtain

$$\eta = \frac{\eta_1 + A}{1 + cnu}\left(1 + \frac{\eta_1 - A}{\eta_1 + A} cnu\right). \tag{6.13}$$

Using (5.7a), (5.15b), (6.1), (6.10a,b), (6.12), (6.13) and general properties of the Jacobian elliptic functions, we obtain after some calculations (see Fig. 5.1c)

$$L^{(1)} = \mathrm{Re}\frac{1}{2}\int_{\Lambda_L} Q(\eta)d\eta = F^{1/2}A^{3/2}(1-\alpha)\mathrm{Re}\frac{1}{2} \times \int_0^{2K+2iK'} \frac{(1-cnu)(k'^2+k^2cn^2u)}{(1+cnu)^2(1+\alpha cnu)}du, \tag{6.14}$$

where $K = K(k)$ and $K' = K(k')$ with

$$k^2 = \frac{A+b_1-\eta_1}{2A}, \quad k'^2 = 1-k^2, \quad \alpha = \frac{\eta_1 - A}{\eta_1 + A}. \tag{6.15a,b,c}$$

Evaluating the integral in (6.14), we obtain

$$L^{(1)} = \frac{F^{1/2}A^{3/2}k^2k'^2}{3(1-\alpha)}\Bigg[-(1+5\alpha)\frac{K(k)-E(k)}{k^2} + (5+\alpha)\frac{E(k)}{k'^2} + \frac{3\alpha(1+\alpha)}{k^2}\Pi(k^2+\alpha^2k'^2,k) - \frac{3\pi(1+\alpha)^{1/2}(k^2+\alpha^2k'^2)^{1/2}}{2(1-\alpha)^{1/2}k^2k'^2}\Bigg]. \tag{6.16}$$

With the aid of the expressions (7.17f) and (7.17f′) in Chapter 7 for β and β' we obtain from (6.16) the formula

$$L^{(1)} = \frac{2F^{1/2}A^{3/2}}{3}\Bigg\{k^2(1-k^2)\left(1+\frac{3\beta}{2}\right)\frac{K(k)-E(k)}{k^2} + k'^2(1-k'^2)\left(1+\frac{3\beta'}{2}\right)\frac{E(k)}{k'^2} - \frac{3\alpha\beta k'^2}{2}\Pi(k^2+\alpha^2k'^2,k) + \frac{3\pi\beta[\beta'(k^2+\alpha^2k'^2)]^{1/2}}{4(\alpha+1)}\Bigg\}. \tag{6.17}$$

Using (7.18a) and (7.21a) in Chapter 7 one sees that (6.17) agrees with (7.22a), with $m \neq 0$, in Chapter 7.

Considering $K^{(1)}$ in the super-barrier case in Fig. 5.1c, one similarly obtains

$$K^{(1)} = -\frac{2F^{1/2}A^{3/2}k^2k'^3}{3(1-\alpha)}\left[(5+\alpha)\frac{K(k')-E(k')}{k'^2} -(1+5\alpha)\frac{E(k')}{k^2} - \frac{3(1+\alpha)}{\alpha k'^2}\Pi\left(\frac{k^2+\alpha^2k'^2}{\alpha^2}, k'\right)\right]. \tag{6.18}$$

With the aid of (6.16b) along with (7.17f) and (7.17d′,e′,f′) in Chapter 7 we obtain from (6.18) the formula

$$K^{(1)} = -\frac{4F^{1/2}A^{3/2}}{3}\left[k'^2(1-k'^2)\left(1+\frac{3\beta'}{2}\right)\frac{K(k')-E(k')}{k'^2} + k^2(1-k^2)\left(1+\frac{3\beta}{2}\right)\frac{E(k')}{k^2} - \frac{3\alpha'\beta'k^2}{2}\Pi(k'^2+k^2\alpha'^2, k')\right]. \tag{6.19}$$

Using (7.18a) and (7.21a) in Chapter 7 one sees that (6.19) agrees with (7.22d), with $m \neq 0$, in Chapter 7.

In Chapter 7 we collect formulas that have been derived as described in the present chapter, although with the use of a computer program.

Adjoined Papers

by Anders Hökback and Per Olof Fröman

Chapter 7

Phase-Inegral Quantities and Their Partial Derivatives with Respect to E and Z_1 Expressed in Terms of Complete Elliptic Integrals

Anders Hökback and Per Olof Fröman

In this chapter we collect and present, without derivation, in explicit, final form the relevant phase-integral quantities and their partial derivatives with respect to E and Z_1 expressed in terms of complete elliptic integrals for the first, third and fifth order of the phase-integral approximation. For the first- and third-order approximations some of the formulas were first derived by means of analytical calculations, and then all formulas were obtained by means of a computer program. In practical calculations it is most convenient to work with real quantities. For the phase-integral quantities associated with the η-equation we therefore give different formulas for the sub-barrier and the super-barrier cases. As in Chapter 6 we use instead of $\tilde{L}_{2n}$, L_{2n}, K_{2n} the notations $\tilde{L}^{(2n+1)}$, $L^{(2n+1)}$, $K^{(2n+1)}$.

It turns out that the $(2n+1)$th-order contributions to the phase-integral quantities needed can be expressed in terms of functions $H^{(2n+1)}(k,\beta)$, $F^{(2n+1)}(k,\beta)$, $G^{(2n+1)}(k,\beta)$ and $\bar{H}^{(2n+1)}(k,\beta)$, $\bar{F}^{(2n+1)}(k,\beta)$, $\bar{G}^{(2n+1)}(k,\beta)$ with various expressions for the parameters k and β. These functions are therefore called "universal" functions. Their explicit expressions are derived for $2n+1$ equal to 1, 3 and 5.

7.1 The ξ-equation

With $\mu = e = \hbar = 1$ we write (5.1) as

$$\tilde{Q}^2(\xi) = -\frac{F(\xi - \xi_0)(\xi - \xi_1)(\xi - \xi_2)}{4\xi^2}, \tag{7.1}$$

where $\xi_0 < 0 < \xi_1 < \xi_2$ when $m \neq 0$ [Fig. 5.1a] but $\xi_0 < 0 = \xi_1 < \xi_2$ when $m = 0$ [Fig. 5.2a]. We then introduce the parameters

$$k^2 = \frac{\xi_2 - \xi_1}{\xi_2 - \xi_0}, \quad \alpha^2 = \frac{\xi_2 - \xi_1}{\xi_2}, \quad \beta = \frac{1}{\alpha^2}, \quad h = (\xi_2 - \xi_0)^{1/2}, \tag{7.2a,b,c,d}$$

$$k'^2 = 1 - k^2, \quad \alpha'^2 = \frac{\alpha^2(1-k^2)}{\alpha^2 - k^2}, \quad \beta' = \frac{1}{\alpha'^2}. \tag{7.2a$'$,b$'$,c$'$}$$

We note that from (7.2a$'$,b$'$) it follows that the expression for k^2 in terms of k'^2 is the same as the expression for k'^2 in terms of k^2, and that the expression for α^2 in terms of k'^2 and α'^2 is the same as the expression for α'^2 in terms of k^2 and α^2. Note also that $\alpha^2 = 1$ for $m = 0$ and that $\alpha'^2 = 1$ when $\alpha^2 = 1$.

We introduce for the first-order approximation the functions

$$h_1(k^2, \beta) = k^2(1-k^2)[3(1-\beta)(1-k^2\beta) - 1], \tag{7.3a}$$

$$f_1(k^2, \beta) = k^2(1-k^2)\beta, \tag{7.3b}$$

$$g_1(k^2, \beta) = 1 - k^2, \tag{7.3c}$$

for the third-order approximation the functions

$$h_3(k^2, \beta) = \frac{3}{1-k^2}[k^2 + 1 + \beta(k^4 - 4k^2 + 1)], \tag{7.4a}$$

$$\begin{aligned} f_3(k^2, \beta) = -\frac{1}{(1-k^2)^3}[&\beta^3(-8k^{10} + 23k^8 - 23k^6 - 23k^4 + 23k^2 - 8) \\ &+ 3\beta^2(4k^8 - 11k^6 + 30k^4 - 11k^2 + 4) \\ &+ \beta(-5k^6 - 19k^4 - 19k^2 - 5) + (5k^4 + 6k^2 + 5)], \end{aligned} \tag{7.4b}$$

$$g_3(k^2,\beta) = -\frac{1}{k^2(1-k^2)^3}[\beta^2(8k^{10} - 23k^8 + 23k^6 + 23k^4 - 23k^2 + 8) + 2\beta(-4k^8 + 11k^6 - 30k^4 + 11k^2 - 4) + (k^6 + 7k^4 + 7k^2 + 1)], \tag{7.4c}$$

and for the fifth-order approximation the functions

$$h_5(k^2,\beta) = \frac{1}{k^4(1-k^2)^5}[7\beta^3(-128k^{18} + 624k^{16} - 1215k^{14} + 1182k^{12} - 591k^{10} - 591k^8 + 1182k^6 - 1215k^4 + 624k^2 - 128) + 6\beta^2(224k^{16} - 1086k^{14} + 2100k^{12} - 2005k^{10} + 2430k^8 - 2005k^6 + 2100k^4 - 1086k^2 + 224) + 3\beta(-168k^{14} + 807k^{12} - 1550k^{10} + 15k^8 + 15k^6 - 1550k^4 + 807k^2 - 168) + 2(14k^{12} - 66k^{10} + 375k^8 + 250k^6 + 375k^4 - 66k^2 + 14)], \tag{7.5a}$$

$$f_5(k^2,\beta) = -\frac{1}{k^4(1-k^2)^7}[7\beta^5(1024k^{24} - 6528k^{22} + 17440k^{20} - 25027k^{18} + 20335k^{16} - 8737k^{14} + 938k^{12} - 8737k^{10} + 20335k^8 - 25027k^6 + 17440k^4 - 6528k^2 + 1024) + 2\beta^4(-8960k^{22} + 57024k^{20} - 152036k^{18} + 218123k^{16} - 178649k^{14} + 82418k^{12} + 82418k^{10} - 178649k^8 + 218123k^6 - 152036k^4 + 57024k^2 - 8960) + 2\beta^3(8064k^{20} - 51216k^{18} + 135841k^{16} - 193367k^{14} + 155635k^{12} - 181594k^{10} + 155635k^8 - 193367k^6 + 135841k^4 - 51216k^2 + 8064) + 8\beta^2(-784k^{18} + 4985k^{16} - 13137k^{14} + 18485k^{12} - 589k^{10} - 589k^8 + 18485k^6 - 13137k^4 + 4985k^2 - 784) + \beta(952k^{16} - 6157k^{14} + 16253k^{12} - 46143k^{10} - 1490k^8 - 46143k^6 + 16253k^4 - 6157k^2 + 952) + 2(-14k^{14} + 97k^{12} + 481k^{10} + 3020k^8 + 3020k^6 + 481k^4 + 97k^2 - 14)], \tag{7.5b}$$

$$
\begin{aligned}
g_5(k^2,\beta) = -\frac{1}{k^6(1-k^2)^7}&[7\beta^4(1024k^{24} - 6528k^{22} + 17440k^{20} \\
&- 25027k^{18} + 20335k^{16} - 8737k^{14} + 938k^{12} - 8737k^{10} \\
&+ 20335k^8 - 25027k^6 + 17440k^4 - 6528k^2 + 1024) \\
&+ 8\beta^3(-1792k^{22} + 11456k^{20} - 30708k^{18} + 44309k^{16} \\
&- 36439k^{14} + 16758k^{12} + 16758k^{10} - 36439k^8 \\
&+ 44309k^6 - 30708k^4 + 11456k^2 - 1792) + 2\beta^2(4480k^{20} \\
&- 28752k^{18} + 77365k^{16} - 112225k^{14} + 92263k^{12} \\
&- 109270k^{10} + 92263k^8 - 112225k^6 + 77365k^4 \\
&- 28752k^2 + 4480) + 8\beta(-224k^{18} + 1446k^{16} - 3905k^{14} \\
&+ 5711k^{12} + 556k^{10} + 556k^8 + 5711k^6 - 3905k^4 + 1446k^2 \\
&- 224) + (56k^{16} - 365k^{14} + 985k^{12} - 5199k^{10} - 5290k^8 \\
&- 5199k^6 + 985k^4 - 365k^2 + 56)]. \qquad (7.5c)
\end{aligned}
$$

For the first-order approximation we introduce the "universal" functions

$$
\begin{aligned}
H^{(1)}(k,\beta) = -\Bigg[&h_1(k'^2,\beta')\frac{K(k)-E(k)}{k^2} + h_1(k^2,\beta)\frac{E(k)}{k'^2} \\
&- 3(1-\delta_{m,0})k^2(1-\beta)(1-k^2\beta)\Pi(\alpha^2,k)\Bigg], \qquad (7.6a)
\end{aligned}
$$

$$
F^{(1)}(k,\beta) = -\left[f_1(k'^2,\beta')\frac{K(k)-E(k)}{k^2} - f_1(k^2,\beta)\frac{E(k)}{k'^2}\right], \qquad (7.6b)
$$

$$
G^{(1)}(k,\beta) = g_1(k'^2,\beta')\frac{K(k)-E(k)}{k^2} + g_1(k^2,\beta)\frac{E(k)}{k'^2}, \qquad (7.6c)
$$

in terms of which we have

$$
\tilde{L}^{(1)} = \frac{F^{1/2}h^3}{3}H^{(1)}(k,\beta), \qquad (7.7a)
$$

$$
\frac{\partial \tilde{L}^{(1)}}{\partial E} = \frac{h}{F^{1/2}}F^{(1)}(k,\beta), \qquad (7.7b)
$$

$$
\frac{\partial \tilde{L}^{(1)}}{\partial Z_1} = \frac{2}{F^{1/2}h}G^{(1)}(k,\beta). \qquad (7.7c)
$$

For the third-order approximation we introduce the "universal" functions

$$H^{(3)}(k,\beta) = h_3(k'^2,\beta')\frac{K(k)-E(k)}{k^2} - h_3(k^2,\beta)\frac{E(k)}{k'^2}, \tag{7.8a}$$

$$F^{(3)}(k,\beta) = f_3(k'^2,\beta')\frac{K(k)-E(k)}{k^2} + f_3(k^2,\beta)\frac{E(k)}{k'^2}, \tag{7.8b}$$

$$G^{(3)}(k,\beta) = -\left[g_3(k'^2,\beta')\frac{K(k)-E(k)}{k^2} + g_3(k^2,\beta)\frac{E(k)}{k'^2}\right], \tag{7.8c}$$

in terms of which we have

$$\tilde{L}^{(3)} = \frac{1}{18F^{1/2}h^3}H^{(3)}(k,\beta), \tag{7.9a}$$

$$\frac{\partial\tilde{L}^{(3)}}{\partial E} = \frac{1}{6F^{3/2}h^5}F^{(3)}(k,\beta), \tag{7.9b}$$

$$\frac{\partial\tilde{L}^{(3)}}{\partial Z_1} = \frac{1}{3F^{3/2}h^7}G^{(3)}(k,\beta). \tag{7.9c}$$

For the fifth-order approximation we introduce the "universal" functions

$$H^{(5)}(k,\beta) = -\left[h_5(k'^2,\beta')\frac{K(k)-E(k)}{k^2} + h_5(k^2,\beta)\frac{E(k)}{k'^2}\right], \tag{7.10a}$$

$$F^{(5)}(k,\beta) = -\left[f_5(k'^2,\beta')\frac{K(k)-E(k)}{k^2} - f_5(k^2,\beta)\frac{E(k)}{k'^2}\right], \tag{7.10b}$$

$$G^{(5)}(k,\beta) = g_5(k'^2,\beta')\frac{K(k)-E(k)}{k^2} + g_5(k^2,\beta)\frac{E(k)}{k'^2}, \tag{7.10c}$$

in terms of which we have

$$\tilde{L}^{(5)} = \frac{1}{360F^{3/2}h^9}H^{(5)}(k,\beta), \tag{7.11a}$$

$$\frac{\partial\tilde{L}^{(5)}}{\partial E} = \frac{1}{120F^{5/2}h^{11}}F^{(5)}(k,\beta), \tag{7.11b}$$

$$\frac{\partial\tilde{L}^{(5)}}{\partial Z_1} = \frac{1}{60F^{5/2}h^{13}}G^{(5)}(k,\beta). \tag{7.11c}$$

7.2 The η-equation in the sub-barrier case

With $\mu = e = \hbar = 1$ we write (5.2) as

$$Q^2(\eta) = \frac{F(\eta-\eta_1)(\eta-\eta_2)(\eta-\eta_3)}{4\eta^2}, \tag{7.12}$$

where $\eta_1 = 0$ when $m = 0$; see Figs. 5.2b and 5.2c. We then introduce the parameters

$$k^2 = \frac{\eta_2-\eta_1}{\eta_3-\eta_1}, \quad \alpha^2 = \frac{\eta_1-\eta_2}{\eta_1}, \quad \beta = \frac{1}{\alpha^2}, \quad h = (\eta_3-\eta_1)^{1/2}, \tag{7.13a,b,c,d}$$

$$k'^2 = 1-k^2, \quad \alpha'^2 = \frac{\alpha^2(1-k^2)}{\alpha^2-k^2}, \quad \beta' = \frac{1}{\alpha'^2}. \tag{7.13a$'$,b$'$,c$'$}$$

We note that from (7.13a$'$,b$'$) it follows that the expression for k^2 in terms of k'^2 is the same as the expression for k'^2 in terms of k^2, and that the expression for α^2 in terms of k'^2 and α'^2 is the same as the expression for α'^2 in terms of k^2 and α^2.

Furthermore, we introduce the same functions $h_{2n+1}(k^2,\beta)$, $f_{2n+1}(k^2,\beta)$ and $g_{2n+1}(k^2,\beta)$, as well as the same "universal" functions $H^{(2n+1)}(k,\beta)$, $F^{(2n+1)}(k,\beta)$ and $G^{(2n+1)}(k,\beta)$ as for the ξ-equation; see Section 7.1. In terms of these "universal" functions we have

$$L^{(1)} = -\frac{F^{1/2}h^3}{3}H^{(1)}(k,\beta), \tag{7.14a}$$

$$\frac{\partial L^{(1)}}{\partial E} = -\frac{h}{F^{1/2}}F^{(1)}(k,\beta), \tag{7.14b}$$

$$\frac{\partial L^{(1)}}{\partial Z_1} = -\frac{2}{F^{1/2}h}G^{(1)}(k,\beta), \tag{7.14c}$$

$$K^{(1)} = \frac{F^{1/2}h^3}{3}H^{(1)}(k',\beta'), \tag{7.14d}$$

$$\frac{\partial K^{(1)}}{\partial E} = -\frac{h}{F^{1/2}}F^{(1)}(k',\beta'), \tag{7.14e}$$

$$\frac{\partial K^{(1)}}{\partial Z_1} = \frac{2}{F^{1/2}h}G^{(1)}(k',\beta'), \tag{7.14f}$$

$$L^{(3)} = -\frac{1}{18F^{1/2}h^3}H^{(3)}(k,\beta), \tag{7.15a}$$

$$\frac{\partial L^{(3)}}{\partial E} = -\frac{1}{6F^{3/2}h^5}F^{(3)}(k,\beta), \tag{7.15b}$$

$$\frac{\partial L^{(3)}}{\partial Z_1} = -\frac{1}{3F^{3/2}h^7}G^{(3)}(k,\beta), \tag{7.15c}$$

$$K^{(3)} = -\frac{1}{18F^{1/2}h^3}H^{(3)}(k',\beta'), \tag{7.15d}$$

$$\frac{\partial K^{(3)}}{\partial E} = \frac{1}{6F^{3/2}h^5}F^{(3)}(k',\beta'), \tag{7.15e}$$

$$\frac{\partial K^{(3)}}{\partial Z_1} = -\frac{1}{3F^{3/2}h^7}G^{(3)}(k',\beta'), \tag{7.15f}$$

$$L^{(5)} = -\frac{1}{360F^{3/2}h^9}H^{(5)}(k,\beta), \tag{7.16a}$$

$$\frac{\partial L^{(5)}}{\partial E} = -\frac{1}{120F^{5/2}h^{11}}F^{(5)}(k,\beta), \tag{7.16b}$$

$$\frac{\partial L^{(5)}}{\partial Z_1} = -\frac{1}{60F^{5/2}h^{13}}G^{(5)}(k,\beta), \tag{7.16c}$$

$$K^{(5)} = \frac{1}{360F^{3/2}h^9}H^{(5)}(k',\beta'), \tag{7.16d}$$

$$\frac{\partial K^{(5)}}{\partial E} = -\frac{1}{120F^{5/2}h^{11}}F^{(5)}(k',\beta'), \tag{7.16e}$$

$$\frac{\partial K^{(5)}}{\partial Z_1} = -\frac{1}{60F^{5/2}h^{13}}G^{(5)}(k',\beta'). \tag{7.16f}$$

7.3 The η-equation in the super-barrier case

Recalling the formula (7.12) for $Q^2(\eta)$, which remains valid in the superbarrier case (still with $\eta_1 = 0$ when $m = 0$), we introduce the parameters [cf. Page 86 in Byrd and Friedman (1971)]

$$a_1 = -\operatorname{Im}\eta_2 = \operatorname{Im}\eta_3(>0), \quad b_1 = \operatorname{Re}\eta_2 = \operatorname{Re}\eta_3, \quad A = [(b_1-\eta_1)^2 + {a_1}^2]^{1/2}, \tag{7.17a,b,c}$$

$$k^2 = \frac{A+b_1-\eta_1}{2A}, \quad \alpha = \frac{\eta_1 - A}{\eta_1 + A}, \quad \beta = \frac{\alpha+1}{\alpha-1}, \tag{7.17d,e,f}$$

$$k'^2 = 1-k^2, \quad \alpha' = 1/\alpha, \quad \beta' = \frac{\alpha'+1}{\alpha'-1} = -\beta. \tag{7.17d',e',f'}$$

We note that from (7.17d′,e′) it follows that the expression for k^2 in terms of k'^2 is the same as the expression for k'^2 in terms of k^2, and that the expression for α in terms of α' is the same as the expression for α' in terms of α.

Furthermore we introduce for the first-order approximation the functions

$$\bar{h}_1(k^2,\beta) = k^2(1-k^2)\left(1+\frac{3}{2}\beta\right), \tag{7.18a}$$

$$\bar{f}_1(k^2,\beta) = k^2(\beta-1), \tag{7.18b}$$

$$\bar{g}_1(k^2,\beta) = k^2, \tag{7.18c}$$

for the third-order approximation the functions

$$\bar{h}_3(k^2,\beta) = 3[4k^2+1+\beta(-32k^4+20k^2+1)], \tag{7.19a}$$

$$\begin{aligned}\bar{f}_3(k^2,\beta) = -\frac{1}{k^2(1-k^2)}&[\beta^3(-4096k^{10}+8704k^8-5632k^6+1028k^4\\ &-5k^2+2)+3\beta^2(512k^8-832k^6+324k^4-7k^2+2)\\ &+\beta(-160k^6+180k^4-23k^2+6)+(4k^4-7k^2+2)],\end{aligned} \tag{7.19b}$$

$$\begin{aligned}\bar{g}_3(k^2,\beta) = -\frac{1}{k^2(1-k^2)}&[\beta^2(-4096k^{10}+8704k^8-5632k^6+1028k^4\\ &-5k^2-2)+2\beta(512k^8-832k^6+324k^4-7k^2+2)\\ &+(-32k^6+36k^4-5k^2+2)],\end{aligned} \tag{7.19c}$$

and for the fifth-order approximation the functions

$$\begin{aligned}\bar{h}_5(k^2,\beta) = -\frac{1}{k^4(1-k^2)^2}&[7\beta^3(-4194304k^{18}+17301504k^{16}\\ &-28311552k^{14}+23138304k^{12}-9694080k^{10}+1878000k^8\\ &-118104k^6+231k^4-3k^2+8)+3\beta^2(3670016k^{16}\\ &-13303808k^{14}+18356224k^{12}-11803904k^{10}+340526k^8\\ &-324792k^6+1037k^4-121k^2+56)+3\beta(-344064k^{14}\\ &+1075200k^{12}-1205760k^{10}+562704k^8-88624k^6\\ &+681k^4-165k^2+56)+(14336k^{12}-37632k^{10}\\ &+32400k^8-9320k^6+285k^4-153k^2+56)],\end{aligned} \tag{7.20a}$$

$$
\begin{aligned}
\bar{f}_5(k^2,\beta) = \frac{1}{k^6(1-k^2)^3}&[7\beta^5(-536870912k^{24} + 3019898880k^{22}\\
&- 7197425664k^{20} + 9427222528k^{18} - 7351107584k^{16}\\
&+ 3446448128k^{14} - 929052672k^{12} + 127330256k^{10}\\
&- 6465440k^8 + 22573k^6 - 149k^4 + 32k^2 + 16)\\
&+ \beta^4(2348810240k^{22} - 12037652480k^{20} + 25618284544k^{18}\\
&- 29121183744k^{16} + 18885009408k^{14} - 6871478784k^{12}\\
&+ 1265964112k^{10} - 88234984k^8 + 484753k^6 - 3393k^4\\
&+ 48k^2 + 560) + 2\beta^3(-264241152k^{20} + 1222115328k^{18}\\
&- 2287730688k^{16} + 2201284608k^{14} - 1135273088k^{12}\\
&+ 292752720k^{10} - 29175440k^8 + 268455k^6 - 783k^4\\
&- 800k^2 + 560) + 2\beta^2(25690112k^{18} - 105971712k^{16}\\
&+ 171026432k^{14} - 134278912k^{12} + 51043120k^{10}\\
&- 7636896k^8 + 127361k^6 + 1639k^4 - 1424k^2 + 560)\\
&+ \beta(-1949696k^{16} + 7067648k^{14} - 9554432k^{12}\\
&+ 5752880k^{10} - 1363600k^8 + 45063k^6 + 3121k^4\\
&- 1824k^2 + 560) + (14336k^{14} - 44800k^{12} + 48336k^{10}\\
&- 19400k^8 + 1245k^6 + 627k^4 - 400k^2 + 112)], \qquad (7.20\text{b})
\end{aligned}
$$

$$
\begin{aligned}
\bar{g}_5(k^2,\beta) = \frac{1}{k^6(1-k^2)^3}&[7\beta^4(-536870912k^{24} + 3019898880k^{22}\\
&- 7197425664k^{20} + 9427222528k^{18} - 7351107584k^{16}\\
&+ 3446448128k^{14} - 929052672k^{12} + 127330256k^{10}\\
&- 6465440k^8 + 22573k^6 - 149k^4 + 32k^2 + 16)\\
&+ 4\beta^3(469762048k^{22} - 2407530496k^{20} + 5122818048k^{18}\\
&- 5820776448k^{16} + 3771392000k^{14} - 1369831936k^{12}\\
&+ 251440784k^{10} - 17363608k^8 + 90621k^6 - 1053k^4\\
&- 16k^2 + 112) + 2\beta^2(-146800640k^{20} + 678952960k^{18}\\
&- 1270349824k^{16} + 1220718592k^{14} - 627705472k^{12}\\
&+ 160824208k^{10} - 15772129k^8 + 134017k^6 - 1618k^4\\
&- 544k^2 + 336) + 4\beta(3670016k^{18} - 15138816k^{16}
\end{aligned}
$$

$$+ 24410112k^{14} - 19113216k^{12} + 7216976k^{10} - 1061048k^8 + 16253k^6 - 61k^4 - 272k^2 + 112) + (-114688k^{16} + 415744k^{14} - 561664k^{12} + 337456k^{10} - 79520k^8 + 2619k^6 + 173k^4 - 288k^2 + 112)]. \tag{7.20c}$$

For the first-order approximation we introduce the "universal" functions

$$\begin{aligned}\bar{H}^{(1)}(k,\beta) &= \bar{h}_1(k^2,\beta)\frac{K(k)-E(k)}{k^2} + \bar{h}_1(k'^2,\beta')\frac{E(k)}{k'^2} \\ &\quad + \frac{3\beta}{2}(1-\delta_{m,0})\left\{\frac{\pi}{2(\alpha+1)}[\beta'(k^2+\alpha^2k'^2)]^{1/2}\right. \\ &\quad \left. - \alpha k'^2\Pi(k^2+\alpha^2k'^2,k)\right\} \\ &\quad + i\left[\bar{h}_1(k'^2,\beta')\frac{K(k')-E(k')}{k'^2} + \bar{h}_1(k^2,\beta)\frac{E(k')}{k^2}\right. \\ &\quad \left. - \frac{3}{2}(1-\delta_{m,0})\alpha'\beta'k^2\Pi(k'^2+\alpha'^2k^2,k')\right],\end{aligned} \tag{7.21a}$$

$$\bar{F}^{(1)}(k,\beta) = -\left[\bar{f}_1(k^2,\beta)\frac{K(k)-E(k)}{k^2} - \bar{f}_1(k'^2,\beta')\frac{E(k)}{k'^2}\right], \tag{7.21b}$$

$$\bar{G}^{(1)}(k,\beta) = -\left[\bar{g}_1(k^2,\beta)\frac{K(k)-E(k)}{k^2} + \bar{g}_1(k'^2,\beta')\frac{E(k)}{k'^2}\right], \tag{7.21c}$$

in terms of which we have

$$L^{(1)} = \frac{2F^{1/2}A^{3/2}}{3}\operatorname{Re}\bar{H}^{(1)}(k,\beta), \tag{7.22a}$$

$$\frac{\partial L^{(1)}}{\partial E} = \frac{A^{1/2}}{2F^{1/2}}\bar{F}^{(1)}(k,\beta), \tag{7.22b}$$

$$\frac{\partial L^{(1)}}{\partial Z_1} = \frac{1}{F^{1/2}A^{1/2}}\bar{G}^{(1)}(k,\beta), \tag{7.22c}$$

$$K^{(1)} = -\frac{4F^{1/2}A^{3/2}}{3}\operatorname{Im}\bar{H}^{(1)}(k,\beta), \tag{7.22d}$$

$$\frac{\partial K^{(1)}}{\partial E} = \frac{A^{1/2}}{F^{1/2}}\bar{F}^{(1)}(k',\beta'), \tag{7.22e}$$

$$\frac{\partial K^{(1)}}{\partial Z_1} = -\frac{2}{F^{1/2}A^{1/2}}\bar{G}^{(1)}(k',\beta'). \tag{7.22f}$$

For the third-order approximation we introduce the "universal" functions

$$\bar{H}^{(3)}(k,\beta) = \bar{h}_3(k^2,\beta)\frac{K(k)-E(k)}{k^2} - \bar{h}_3(k'^2,\beta')\frac{K(k)}{k'^2} - i\left[\bar{h}_3(k'^2,\beta')\frac{K(k')-E(k')}{k'^2} - \bar{h}_3(k^2,\beta)\frac{K(k')}{k^2}\right], \tag{7.23a}$$

$$\bar{F}^{(3)}(k,\beta) = -\left[\bar{f}_3(k^2,\beta)\frac{K(k)-E(k)}{k^2} + \bar{f}_3(k'^2,\beta')\frac{K(k)}{k'^2}\right], \tag{7.23b}$$

$$\bar{G}^{(3)}(k,\beta) = -\left[\bar{g}_3(k^2,\beta)\frac{K(k)-E(k)}{k^2} - \bar{g}_3(k'^2,\beta')\frac{K(k)}{k'^2}\right], \tag{7.23c}$$

in terms of which we have

$$L^{(3)} = \frac{1}{144F^{1/2}A^{3/2}}\operatorname{Re}\bar{H}^{(3)}(k,\beta), \tag{7.24a}$$

$$\frac{\partial L^{(3)}}{\partial E} = \frac{1}{192F^{3/2}A^{5/2}}\bar{F}^{(3)}(k,\beta), \tag{7.24b}$$

$$\frac{\partial L^{(3)}}{\partial Z_1} = \frac{1}{96F^{3/2}A^{7/2}}\bar{G}^{(3)}(k,\beta), \tag{7.24c}$$

$$K^{(3)} = -\frac{1}{72F^{1/2}A^{3/2}}\operatorname{Im}\bar{H}^{(3)}(k,\beta), \tag{7.24d}$$

$$\frac{\partial K^{(3)}}{\partial E} = -\frac{1}{96F^{3/2}A^{5/2}}\bar{F}^{(3)}(k',\beta'), \tag{7.24e}$$

$$\frac{\partial K^{(3)}}{\partial Z_1} = \frac{1}{48F^{3/2}A^{7/2}}\bar{G}^{(3)}(k',\beta'). \tag{7.24f}$$

For the fifth-order approximation we introduce the "universal" functions

$$\bar{H}^{(5)}(k,\beta) = \bar{h}_5(k^2,\beta)\frac{K(k)-E(k)}{k^2} + \bar{h}_5(k'^2,\beta')\frac{K(k)}{k'^2} + i\left[\bar{h}_5(k'^2,\beta')\frac{K(k')-E(k')}{k'^2} + \bar{h}_5(k^2,\beta)\frac{K(k')}{k^2}\right], \tag{7.25a}$$

$$\bar{F}^{(5)}(k,\beta) = -\left[\bar{f}_5(k^2,\beta)\frac{K(k)-E(k)}{k^2} - \bar{f}_5(k'^2,\beta')\frac{K(k)}{k'^2}\right], \tag{7.25b}$$

$$\bar{G}^{(5)}(k,\beta) = -\left[\bar{g}_5(k^2,\beta)\frac{K(k)-E(k)}{k^2} + \bar{g}_5(k'^2,\beta')\frac{K(k)}{k'^2}\right], \tag{7.25c}$$

in terms of which we have

$$L^{(5)} = \frac{1}{46080F^{3/2}A^{9/2}} \mathrm{Re}\, \bar{H}^{(5)}(k, \beta), \tag{7.26a}$$

$$\frac{\partial L^{(5)}}{\partial E} = \frac{1}{61440F^{5/2}A^{11/2}} \bar{F}^{(5)}(k, \beta), \tag{7.26b}$$

$$\frac{\partial L^{(5)}}{\partial Z_1} = \frac{1}{30720F^{5/2}A^{13/2}} \bar{G}^{(5)}(k, \beta). \tag{7.26c}$$

$$K^{(5)} = -\frac{1}{23040F^{3/2}A^{9/2}} \mathrm{Im}\, \bar{H}^{(5)}(k, \beta), \tag{7.26d}$$

$$\frac{\partial K^{(5)}}{\partial E} = \frac{1}{30720F^{5/2}A^{11/2}} \bar{F}^{(5)}(k', \beta'). \tag{7.26e}$$

$$\frac{\partial K^{(5)}}{\partial Z_1} = -\frac{1}{15360F^{5/2}A^{13/2}} \bar{G}^{(5)}(k', \beta'). \tag{7.26f}$$

Chapter 8

Numerical Results

Anders Hökback and Per Olof Fröman

Values of the energy E and the half-width Γ for different states of a hydrogen atom in an electric field $F[= \overline{F}$ according to (2.17)] of various strengths, obtained both in previous work by other authors and in the present work with the use of the phase-integral formulas, are presented in the tables of the present chapter. We use atomic units (au), i.e., such units that $\mu = e = \hbar = 1$. The positions E of the Stark levels were obtained from (5.33) except for the state with $n = 30$ in Table 8.7, where the more accurate formula (5.40) along with (5.42) has been used. The half-widths Γ were obtained from (5.54) along with (5.55) and are therefore accurate only when the barrier is sufficiently thick, which means that Γ is sufficiently small.

We emphasize that the results obtained by us, as well as those obtained by the other authors quoted in this chapter, are obtained by neglecting the fine structure corrections. This is not a serious disadvantage for us, since our main intention has been to compare the accuracy obtainable by the phase-integral method with the accuracy obtainable by other methods of computation. For the experimental data corresponding to the theoretical values presented in this chapter we refer to the publications mentioned in this chapter.

Table 8.1 gives results concerning the Stark effect for the ground state ($m = n_1 = n_2 = 0, n = 1$) of the hydrogen atom. For each value of the field strength F, the first line gives results obtained by Hehenberger, McIntosh and Brändas (1974) by means of Weyl's

Table 8.1. $m = n_1 = n_2 = 0, n = 1$.

F	$-E$	Γ	K	Z_1
0.04	0.503 771 8	3.9×10^{-6}		
	0.503 771 591	3.89×10^{-6}		
	0.505 4	3.1×10^{-6}	5.38	0.516 8
	0.503 87	3.97×10^{-6}	5.258	0.520 26
	0.503 74	3.83×10^{-6}	5.275 7	0.520 225
	0.501 71	3.91×10^{-6}	5.27	0.515 0
	0.503 67	4.02×10^{-6}	5.252	0.520 17
	0.503 76	3.83×10^{-6}	5.276 1	0.520 232
0.08	0.517 56	4.54×10^{-3}		
	0.517 495 363	$4.511 0 \times 10^{-3}$		
	0.519 1	3.9×10^{-3}	1.73	0.535 8
	0.517 9	4.63×10^{-3}	1.639	0.542 5
	0.517 3	4.48×10^{-3}	1.655	0.542 30
	0.507	5.9×10^{-3}	1.54	0.530
	0.515 6	5.0×10^{-3}	1.60	0.541 4
	0.517 7	4.3×10^{-3}	1.66	0.542 5
0.12	0.537 4	2.99×10^{-2}		
	0.535 567	$2.942 3 \times 10^{-2}$		
	0.536 51	2.83×10^{-2}	0.68	0.554 85
	0.536 40	3.26×10^{-2}	0.603	0.564 34
	0.535 21	3.15×10^{-2}	0.620	0.564 02
	0.520	3.6×10^{-2}	0.49	0.548
	0.538	2.4×10^{-2}	0.63	0.565
	0.551	1.4×10^{-2}	0.79	0.570
0.16	0.555 4	7.14×10^{-2}		
	0.547 78	$7.119 5 \times 10^{-2}$		
	0.548 5	7.8×10^{-2}	0.17	0.570 4
	0.549 0	9.0×10^{-2}	0.10	0.582 1
	0.547 3	8.6×10^{-2}	0.12	0.581 7
	0.537	5.2×10^{-2}	0.07	0.566
	0.568	3.0×10^{-2}	0.27	0.590
	0.595	1.3×10^{-2}	0.55	0.600
0.20	0.570 55	1.209×10^{-1}		
	0.552 60	$1.249 3 \times 10^{-1}$		
	0.553 5	1.5×10^{-1}	−0.15	0.582 3
	0.554 2	1.80×10^{-1}	−0.22	0.595 8
	0.552 2	1.71×10^{-1}	−0.20	0.595 4
	0.56	0.9×10^{-1}	−0.11	0.584
	0.51	1.2×10^{-1}	−0.52	0.580
	0.48	1.1×10^{-1}	−0.73	0.567

(Continued)

Table 8.1. (*Continued*)

F	$-E$	Γ	K	Z_1
0.24	0.583	1.74×10^{-1}		
	0.550 82	$1.892\,7 \times 10^{-1}$		
	0.552 3	2.6×10^{-1}	−0.38	0.591
	0.552 7	3.1×10^{-1}	−0.46	0.606 2
	0.550 9	2.9×10^{-1}	−0.43	0.605 9
	0.57	1.8×10^{-1}	−0.3	0.597
	0.52	2.4×10^{-1}	−0.7	0.594
	0.49	2.0×10^{-1}	−0.8	0.582
0.28	–	–		
	0.543 4	2.643×10^{-1}		
	0.545 8	4.0×10^{-1}	−0.56	0.598
	0.545 4	4.8×10^{-1}	−0.64	0.613 8
	0.544 3	4.5×10^{-1}	−0.61	0.614 0
	0.57	3.0×10^{-1}	−0.4	0.605 5
	0.52	4.0×10^{-1}	−0.8	0.604 7
	0.49	3.4×10^{-1}	−0.9	0.594
0.32	–	–		
	0.531 1	3.507×10^{-1}		
	0.534 8	5.7×10^{-1}	−0.71	0.602
	0.533 1	7.0×10^{-1}	−0.80	0.619 3
	0.533 4	6.5×10^{-1}	−0.75	0.620 1
	0.56	4.6×10^{-1}	−0.6	0.611 8
	0.51	6.1×10^{-1}	−0.9	0.612 1
	0.48	5.3×10^{-1}	−1.0	0.602
0.36	–	–		
	0.514 4	4.497×10^{-1}		
	0.519 9	7.8×10^{-1}	−0.83	0.605
	0.516 1	9.8×10^{-1}	−0.93	0.622 8
	0.518 6	9.1×10^{-1}	−0.88	0.624 5
	0.55	6.5×10^{-1}	−0.7	0.616 2
	0.50	8.8×10^{-1}	−1.0	0.617 1
	0.47	7.7×10^{-1}	−1.1	0.609
0.40	–	–		
	0.493 8	0.563 1		
	0.502	1.0	−0.9	0.607
	0.495	1.3	−1.1	0.624 7
	0.501	1.2	−1.0	0.627 6
	0.54	0.9	−0.8	0.619 0
	0.48	1.2	−1.1	0.620 2
	0.46	1.1	−1.2	0.614

theory, the second line gives numerical results obtained by Damburg and Kolosov (1976a), and the lines 3–8 give our phase-integral results both with $\tilde{\phi}$ included (the lines 3, 4 and 5 giving the results in the first-, third- and fifth-order approximations, respectively) and with $\tilde{\phi}$ neglected (the lines 6, 7 and 8 giving the results in the first-, third- and fifth-order approximations, respectively).

Damburg and Kolosov (1976a) state that their data for E are in complete agreement with those of Alexander (1969).

By means of a variational method Froelich and Brändas (1975) have calculated values of E which are close to those given by Hehenberger, McIntosh and Brändas (1974).

Korsch and Möhlenkamp (1983) have used a semiclassical complex-energy treatment to calculate results corresponding to those obtained by Hehenberger, McIntosh and Brändas (1974). They get results that agree approximately with those obtained by Hehenberger, McIntosh and Brändas (1974) although with a less number of digits.

Tables 8.2(a), (b), (c) give results concerning the Stark effect for the three states $m = 0, n_1 = 0, n_2 = 1$ (Table 8.2(a)), $|m| = 1, n_1 = 0, n_2 = 0$ (Table 8.2(b)), and $m = 0, n_1 = 1, n_2 = 0$ (Table 8.2(c)) of the hydrogen atom with the principal quantum number $n = |m| + 1 + n_1 + n_2 = 2$ in an electric field. For each state and each value of the electric field strength F, the first line gives the value of E calculated by the use of the perturbation expansion including terms up to the fourth power of F (Alliluev and Malkin (1974)) given by Damburg and Kolosov (1976a) in their Table 2, the second line gives the numerical results for E and Γ obtained by Damburg and Kolosov (1976a), and the lines 3–8 give our phase-integral results both with $\tilde{\phi}$ included (the lines 3, 4 and 5 giving the results in the first-, third- and fifth-order approximations, respectively) and with $\tilde{\phi}$ neglected (the lines 6, 7 and 8 giving the results in the first-, third- and fifth-order approximations, respectively).

The agreement of our optimum phase-integral results for E with the numerical results obtained by Damburg and Kolosov (1976a) is in general better for the state $m = 0, n_1 = 0, n_2 = 1$ (Table 8.2(a)) than for the states $|m| = 1, n_1 = 0, n_2 = 0$ (Table 8.2(b)) and $m = 0, n_1 = 1, n_2 = 0$ (Table 8.2(c)).

Table 8.2(a). $m = 0, n_1 = 0, n_2 = 1, n = 2.$

F	$-E$	Γ	K	Z_1
0.004	0.138 509 8			
	0.138 548 793	$4.439\,3 \times 10^{-6}$		
	0.138 69	4.1×10^{-6}	4.14	0.268 5
	0.138 551	4.443×10^{-6}	4.101 4	0.270 021
	0.138 548 72	$4.438\,9 \times 10^{-6}$	4.101 9	0.270 023 4
	0.138 14	4.9×10^{-6}	4.06	0.268 0
	0.138 523	4.49×10^{-6}	4.097	0.270 00
	0.138 543	4.450×10^{-6}	4.101 1	0.270 019
0.008	0.156 230 9			
	0.156 376 8	$4.161\,6 \times 10^{-3}$		
	0.156 45	4.5×10^{-3}	0.43	0.289
	0.156 387	$4.673\,2 \times 10^{-3}$	0.404 0	0.291 315
	0.156 376 4	$4.670\,6 \times 10^{-3}$	0.404 3	0.291 322 8
	0.154 9	4.7×10^{-3}	0.30	0.287
	0.157 4	2.7×10^{-3}	0.49	0.292
	0.159	1.3×10^{-3}	0.67	0.294
0.012	0.181 138			
	0.171 517	$1.899\,0 \times 10^{-2}$		
	0.171 62	3.0×10^{-2}	−0.61	0.305
	0.171 536	3.105×10^{-2}	−0.633 5	0.308 47
	0.171 517 2	3.103×10^{-2}	−0.633 1	0.308 484
	0.174	2.4×10^{-2}	−0.49	0.306 6
	0.170 1	2.7×10^{-2}	−0.72	0.307 4
	0.168	2.4×10^{-2}	−0.86	0.305 5
0.016	0.217 79			
	0.179 94	$0.441\,19 \times 10^{-1}$		
	0.180 2	0.96×10^{-1}	−1.27	0.315
	0.179 964	1.015×10^{-1}	−1.303 2	0.319 36
	0.179 939 6	1.014×10^{-1}	−1.302 6	0.319 387
	0.183	0.89×10^{-1}	−1.1	0.317 5
	0.179 930	0.99×10^{-1}	−1.305	0.319 33
	0.178 7	0.96×10^{-1}	−1.36	0.318 5
0.020	0.272 3			
	0.181 1	$0.873\,9 \times 10^{-1}$		
	0.181 9	2.5×10^{-1}	−1.87	0.320
	0.181 165	2.742×10^{-1}	−1.910 2	0.324 42
	0.181 140	2.740×10^{-1}	−1.909 3	0.324 478
	0.185	2.4×10^{-1}	−1.73	0.323
	0.181 6	2.73×10^{-1}	−1.89	0.324 7
	0.180 87	2.71×10^{-1}	−1.92	0.324 28

Table 8.2(b). $|m| = 1, n_1 = 0, n_2 = 0, n = 2.$

F	$-E$	Γ	K	Z_1
0.004	0.126 304 8			
	0.126 316 885	8.1×10^{-7}		
	0.126 43	7.2×10^{-7}	5.05	0.523 0
	0.126 323	8.19×10^{-7}	4.980	0.524 364
	0.126 315 3	8.028×10^{-7}	4.989 4	0.524 354
	0.126 0	8.35×10^{-7}	4.975	0.522
	0.126 302	8.25×10^{-7}	4.976	0.524 33
	0.126 314 6	8.032×10^{-7}	4.989 3	0.524 353
0.008	0.130 901 1			
	0.131 188 59	$2.024\,4 \times 10^{-3}$		
	0.131 16	1.97×10^{-3}	0.93	0.549
	0.131 22	2.14×10^{-3}	0.879	0.551 55
	0.131 175	2.10×10^{-3}	0.889	0.551 48
	0.129	2.7×10^{-3}	0.76	0.546
	0.131 177	1.96×10^{-3}	0.875	0.551 47
	0.132 1	1.4×10^{-3}	0.97	0.553 1
0.012	0.140 834 4			
	0.135 971 6	$1.168\,7 \times 10^{-2}$		
	0.135 90	1.47×10^{-2}	0.18	0.572
	0.136 11	1.59×10^{-2}	0.23	0.575 82
	0.136 05	1.55×10^{-2}	0.21	0.575 75
	0.136 7	1.0×10^{-2}	0.13	0.574
	0.132	1.1×10^{-2}	0.46	0.570
	0.129	0.9×10^{-2}	0.67	0.564
0.016	0.159 514			
	0.136 437	$2.755\,2 \times 10^{-2}$		
	0.136 91	4.4×10^{-2}	−0.79	0.588
	0.137 00	4.9×10^{-2}	−0.84	0.591 8
	0.137 15	4.7×10^{-2}	−0.82	0.592 1
	0.139	3.9×10^{-2}	−0.66	0.592
	0.135 9	4.4×10^{-2}	−0.90	0.590
	0.134	3.9×10^{-2}	−0.99	0.587
0.020	0.191 71			
	0.131 50	$0.518\,81 \times 10^{-1}$		
	0.134 1	1.0×10^{-1}	−1.24	0.597
	0.133 7	1.10×10^{-1}	−1.32	0.600 4
	0.134 3	1.08×10^{-1}	−1.29	0.601 4
	0.137	0.9×10^{-1}	−1.11	0.601 3
	0.133 6	1.07×10^{-1}	−1.33	0.600 2
	0.132 8	1.02×10^{-1}	−1.35	0.599 0

Table 8.2(c). $m = 0, n_1 = 1, n_2 = 0, n = 2.$

F	$-E$	Γ	K	Z_1
0.004	0.114 310 2			
	0.114 305 339	$1.364\,3 \times 10^{-7}$		
	0.114 40	1.2×10^{-7}	5.99	0.769 4
	0.114 314	1.39×10^{-7}	5.893	0.770 61
	0.114 302	$1.340\,3 \times 10^{-7}$	5.912 4	0.770 577
	0.113 99	1.32×10^{-7}	5.92	0.768 2
	0.114 295	1.40×10^{-7}	5.890	0.770 55
	0.114 304 3	$1.339\,7 \times 10^{-7}$	5.912 7	0.770 583
0.008	0.106 633 5			
	0.106 668 4	$8.509\,6 \times 10^{-4}$		
	0.106 58	7.7×10^{-4}	1.48	0.791 7
	0.106 72	8.8×10^{-4}	1.40	0.794 52
	0.106 645	8.49×10^{-4}	1.42	0.794 35
	0.105 0	10.7×10^{-4}	1.33	0.788
	0.106 4	9.2×10^{-4}	1.38	0.793 7
	0.106 80	7.8×10^{-4}	1.44	0.794 7
0.012	0.103 747			
	0.100 621	$6.881\,5 \times 10^{-3}$		
	0.100 3	7.4×10^{-3}	0.24	0.814
	0.100 72	8.3×10^{-3}	0.173	0.818 5
	0.100 58	7.9×10^{-3}	0.197	0.818 2
	0.098 8	6.3×10^{-3}	0.14	0.811
	0.103	3.7×10^{-3}	0.30	0.823
	0.106	1.6×10^{-3}	0.55	0.831
0.016	0.109 013			
	0.092 728	1.7147×10^{-2}		
	0.092 3	2.3×10^{-2}	−0.37	0.830
	0.092 83	2.60×10^{-2}	−0.445	0.835 2
	0.092 76	2.45×10^{-2}	−0.411	0.835 04
	0.094	1.7×10^{-2}	−0.28	0.834
	0.090	2.0×10^{-2}	−0.61	0.828
	0.086	1.6×10^{-2}	−0.77	0.819
0.020	0.127 38			
	0.082 41	3.044×10^{-2}		
	0.082 0	4.9×10^{-2}	−0.79	0.839
	0.082 39	5.7×10^{-2}	−0.88	0.844 2
	0.082 73	5.3×10^{-2}	−0.83	0.844 98
	0.085	4.2×10^{-2}	−0.66	0.845
	0.081	5.1×10^{-2}	−0.93	0.841
	0.079	4.4×10^{-2}	−1.0	0.836

Tables 8.3(a), (b) give results concerning the Stark effect for the two states $m = 0, n_1 = 0, n_2 = 4$ (Table 8.3(a)) and $m = 0, n_1 = 4, n_2 = 0$ (Table 8.3(b)) of a hydrogen atom with the principal quantum number $n = |m| + 1 + n_1 + n_2 = 5$. For each value of the electric field strength F, the first line gives the value of E calculated by the use of the perturbation expansion including terms up to the fourth power of F (Alliluev and Melkin (1974)) given by Damburg and Kolosov (1976a) in their Table 3, the second line gives the numerical results for E and Γ obtained by Damburg and Kolosov (1976a), the third line gives the numerical results for E and Γ calculated by Luc-Koenig and Bachelier (1980a) in their Table 2, and the lines 4–9 give our phase-integral results both with $\tilde{\phi}$ included (the lines 4, 5 and 6 giving the results in the first-, third- and fifth-order approximations, respectively) and with $\tilde{\phi}$ neglected (the lines 7, 8 and 9 giving the results in the first-, third- and fifth-order approximations, respectively).

The agreement of our optimum phase-integral results for E with the numerical results obtained by Damburg and Kolosov (1976a) and by Luc-Koenig and Bachelier (1980(a)) is very good for the state $m = 0, n_1 = 0, n_2 = 4$ (Table 8.3(a)) but in general less good for the state $m = 0, n_1 = 4, n_2 = 0$ (Table 8.3(b)).

Table 8.4 gives results concerning the Stark effect for the state $|m| = 1, n_1 = 3, n_2 = 0$ of the hydrogen atom with the principal quantum number $n = |m| + 1 + n_1 + n_2 = 5$. For each value of the electric field strength F the first line gives the results of Guschina and Nikulin (1975) quoted by Damburg and Kolosov (1976a) in their Table 4, the second line gives the numerical results obtained by Damburg and Kolosov (1976a) in their Table 4, the third line gives the numerical results obtained by Luc-Koenig and Bachelier (1980a) in their Table 2, and the lines 4–9 give our phase-integral results both with $\tilde{\phi}$ included (the lines 4, 5 and 6 giving the results in the first-, third- and fifth-order approximations, respectively) and with $\tilde{\phi}$ neglected (the lines 7, 8 and 9 giving the results in the first-, third- and fifth-order approximations, respectively).

Although the results obtained by Guschina and Nikulin (1975) and by Damburg and Kolosov (1976a) are rather close to each other, the latter authors believe that their data are more accurate. Comparison with our results is somewhat difficult, since we do not know

Table 8.3(a). $m = 0, n_1 = 0, n_2 = 4, n = 5.$

$F \times 10^4$	$-E$	Γ	K	Z_1
1.0	0.023 177 14			
	0.023 179 196 2	4.2×10^{-12}		
	0.023 179 196 289 73	4.229×10^{-12}		
	0.023 183	4.10×10^{-12}	9.67	0.108 46
	0.023 179 21	$4.227\,9 \times 10^{-12}$	9.650 9	0.108 710 26
	0.023 179 196 23	$4.227\,8 \times 10^{-12}$	9.650 9	0.108 710 362
	0.023 169	4.39×10^{-12}	9.63	0.108 43
	0.023 179 10	4.230×10^{-12}	9.650 7	0.108 710 0
	0.023 179 193	$4.227\,9 \times 10^{-12}$	9.650 9	0.108 710 355
1.5	0.024 932 17			
	0.024 956 749	1.919×10^{-6}		
	0.024 956 749 516	1.922×10^{-6}		
	0.024 960	1.88×10^{-6}	3.05	0.112 8
	0.024 956 77	$1.920\,2 \times 10^{-6}$	3.038 0	0.113 166 3
	0.024 956 749 42	$1.920\,2 \times 10^{-6}$	3.038 0	0.113 166 54
	0.024 92	2.18×10^{-6}	2.98	0.112 7
	0.024 954	1.95×10^{-6}	3.033	0.113 159
	0.024 955 9	1.928×10^{-6}	3.036 6	0.113 165
2.0	0.026 837 90			
	0.026 971 36	$1.783\,0 \times 10^{-4}$		
	—	—		
	0.026 973	1.92×10^{-4}	0.517	0.117 5
	0.026 971 40	$1.947\,7 \times 10^{-4}$	0.508 90	0.117 919 4
	0.026 971 367 0	$1.947\,7 \times 10^{-4}$	0.508 90	0.117 919 84
	0.026 89	2.2×10^{-4}	0.40	0.117 3
	0.027 02	1.3×10^{-4}	0.57	0.118 01
	0.027 1	0.7×10^{-4}	0.7	0.118 2
2.5	0.028 933 41			
	0.028 968 28	$0.853\,10 \times 10^{-3}$		
	—	—		
	0.028 970	1.369×10^{-3}	−0.713	0.121 9
	0.028 968 34	$1.383\,0 \times 10^{-3}$	−0.720 31	0.122 426 9
	0.028 968 293	$1.383\,0 \times 10^{-3}$	−0.720 31	0.122 427 47
	0.029 1	1.21×10^{-3}	−0.61	0.122 1
	0.028 93	1.25×10^{-3}	−0.76	0.122 35
	0.028 8	1.12×10^{-3}	−0.87	0.122 2

(*Continued*)

Table 8.3(a). *(Continued)*

$F \times 10^4$	$-E$	Γ	K	Z_1
3.0	0.031 265 1			
	0.030 538 1	$1.969\,8 \times 10^{-3}$		
	—	—		
	0.030 543	5.07×10^{-3}	−1.728	0.125 4
	0.030 538 16	$5.123\,3 \times 10^{-3}$	−1.737 9	0.125 922 8
	0.030 538 093	$5.123\,3 \times 10^{-3}$	−1.737 9	0.125 923 60
	0.030 65	4.90×10^{-3}	−1.63	0.125 6
	0.030 55	5.10×10^{-3}	−1.726	0.125 95
	0.030 528	5.06×10^{-3}	−1.748	0.125 90
3.5	0.033 887 0			
	0.031 433 8	$0.364\,34 \times 10^{-2}$		
	—	—		
	0.031 447	1.61×10^{-2}	−2.88	0.127 4
	0.031 434 02	$1.632\,8 \times 10^{-2}$	−2.900 2	0.128 024 1
	0.031 433 939	$1.632\,8 \times 10^{-2}$	−2.900 2	0.128 025 14
	0.031 55	1.56×10^{-2}	−2.80	0.127 6
	0.031 444	$1.632\,4 \times 10^{-2}$	−2.891	0.128 044
	0.031 437	$1.632\,4 \times 10^{-2}$	−2.898	0.128 030
4.0	0.036 860			
	0.031 408	$0.635\,13 \times 10^{-2}$		
	—	—		
	0.031 44	5.67×10^{-2}	−4.36	0.127 7
	0.031 408 71	$5.801\,5 \times 10^{-2}$	−4.384 1	0.128 336 6
	0.031 408 603	$5.801\,6 \times 10^{-2}$	−4.384 1	0.128 338 12
	0.031 54	5.47×10^{-2}	−4.28	0.127 9
	0.031 413	5.799×10^{-2}	−4.381	0.128 345
	0.031 409 7	$5.801\,5 \times 10^{-2}$	−4.383 3	0.128 340
4.5	0.040 25			
	0.029 98	$0.127\,3 \times 10^{-1}$		
	—	—		
	0.030 06	3.04×10^{-1}	−6.38	0.125 3
	0.029 984 6	$3.184\,3 \times 10^{-1}$	−6.428 8	0.125 966
	0.029 984 40	$3.184\,6 \times 10^{-1}$	−6.428 8	0.125 968 3
	0.030 18	2.90×10^{-1}	−6.30	0.125 5
	0.029 987	$3.182\,8 \times 10^{-1}$	−6.427 5	0.125 970
	0.029 984 6	$3.184\,5 \times 10^{-1}$	−6.428 7	0.125 968 8

Table 8.3(b). $m = 0, n_1 = 4, n_2 = 0, n = 5.$

$F \times 10^4$	$-E$	Γ	K	Z_1
1.5	0.015 813 91			
	0.015 807 764 5	2×10^{-11}		
	0.015 807 764 455	1.432×10^{-11}		
	0.015 808 3	1.2×10^{-11}	9.18	0.913 3
	0.015 808 1	1.464×10^{-11}	9.099 7	0.913 640
	0.015 807 66	$1.406 3 \times 10^{-11}$	9.119 6	0.913 632
	0.015 79	1.3×10^{-11}	9.15	0.913 0
	0.015 807 58	1.466×10^{-11}	9.098 8	0.913 631
	0.015 807 757	$1.405 9 \times 10^{-11}$	9.119 7	0.913 634 3
2.0	0.014 557 59			
	0.014 535 204 9	4.026×10^{-8}		
	0.014 535 205 2	4.029×10^{-8}		
	0.014 532	3.5×10^{-8}	5.18	0.918 2
	0.014 535 8	4.12×10^{-8}	5.101	0.918 61
	0.014 534 97	$3.952 8 \times 10^{-8}$	5.120 9	0.918 596
	0.014 50	3.9×10^{-8}	5.14	0.917 6
	0.014 534 2	4.14×10^{-8}	5.098	0.918 57
	0.014 535 07	$3.953 0 \times 10^{-8}$	5.121 1	0.918 591
2.5	0.013 385 92			
	0.013 328 925	$3.271 9 \times 10^{-6}$		
	—	—		
	0.013 320	2.9×10^{-6}	2.93	0.923 4
	0.013 330 2	3.35×10^{-6}	2.859	0.924 02
	0.013 328 4	3.21×10^{-6}	2.879 1	0.923 987
	0.013 27	3.35×10^{-6}	2.87	0.922 4
	0.013 323	3.43×10^{-6}	2.850	0.923 89
	0.013 327	3.23×10^{-6}	2.878	0.923 97
3.0	0.012 319 09			
	0.012 200 93	$4.166 6 \times 10^{-5}$		
	—	—		
	0.012 18	3.8×10^{-5}	1.58	0.929 3
	0.012 203	4.3×10^{-5}	1.510	0.930 16
	0.012 199 6	4.14×10^{-5}	1.531	0.930 086
	0.012 09	4.8×10^{-5}	1.48	0.928
	0.012 186	4.5×10^{-5}	1.49	0.929 8
	0.012 206	4.0×10^{-5}	1.537	0.930 20
3.5	0.011 384 69			
	0.011 136 04	$1.791 4 \times 10^{-4}$		
	—	—		
	0.011 11	1.73×10^{-4}	0.73	0.935 6

(Continued)

Table 8.3(b). *(Continued)*

$F \times 10^4$	$-E$	Γ	K	Z_1
	0.011 140	1.96×10^{-4}	0.660	0.936 75
	0.011 133 8	1.87×10^{-4}	0.683	0.936 634
	0.011 0	2.1×10^{-4}	0.60	0.933
	0.011 15	1.67×10^{-4}	0.67	0.936 9
	0.011 24	1.1×10^{-4}	0.79	0.939
4.0	0.010 617 72			
	0.010 082 06	$4.280 4 \times 10^{-4}$		
	–	–		
	0.010 04	4.6×10^{-4}	0.14	0.941 5
	0.010 087	5.2×10^{-4}	0.08	0.942 88
	0.010 080 1	5.0×10^{-4}	0.10	0.942 756
	0.010 0	3.6×10^{-4}	0.08	0.940
	0.010 3	2.1×10^{-4}	0.2	0.946
	0.010 5	0.9×10^{-4}	0.5	0.950
4.5	0.010 060 59			
	0.008 994 79	$7.588 2 \times 10^{-4}$		
	–	–		
	0.008 95	9.5×10^{-4}	−0.30	0.946 4
	0.008 999	10.8×10^{-4}	−0.37	0.947 83
	0.008 996 4	10.2×10^{-4}	−0.34	0.947 796
	0.009 03	7.3×10^{-4}	−0.2	0.947 8
	0.008 8	7.9×10^{-4}	−0.5	0.945
	0.008 6	6.1×10^{-4}	−0.7	0.941
5.0	0.009 763 1			
	0.007 851 7	$1.148 3 \times 10^{-3}$		
	–	–		
	0.007 81	1.7×10^{-3}	−0.66	0.950 0
	0.007 850 9	1.9×10^{-3}	−0.75	0.951 3
	0.007 862 2	1.8×10^{-3}	−0.70	0.951 5
	0.007 9	1.5×10^{-3}	−0.6	0.952
	0.007 77	1.7×10^{-3}	−0.8	0.950
	0.007 6	1.4×10^{-3}	−0.9	0.948

exactly for which values of F (expressed in atomic units) the other authors have performed their calculations. Their field strengths were originally expressed in V/cm, and we do not know the exact values of the conversion factors used by these authors to express the field strengths, originally given in V/cm, in au. In our phase-integral calculations the values of F in atomic units (given in our Table 8.4) were considered to be exact.

Table 8.4. $|m| = 1, n_1 = 3, n_2 = 0, n = 5$.

F	$-E$	Γ	K	Z_1
1.556×10^{-4} au (0.8×10^6 V/cm)	0.016 840 127 3	5.03×10^{-10}		0.824 056 67
	0.016 855 237 2	4.2×10^{-10}		
	0.016 855 237 14	4.222×10^{-10}		
	0.016 856 0	3.8×10^{-10}	7.45	0.823 7
	0.016 855 4	4.27×10^{-10}	7.392	0.824 022
	0.016 855 18	4.180×10^{-10}	7.402 3	0.824 017 4
	0.016 836	4.1×10^{-10}	7.41	0.823 4
	0.016 854 9	4.28×10^{-10}	7.391	0.824 013
	0.016 855 219	4.179×10^{-10}	7.402 4	0.824 018 1
$1.944 8 \times 10^{-4}$ au (1.0×10^6 V/cm)	0.016 162 71	1.48×10^{-7}		0.830 52
	0.016 179 388 5	1.438×10^{-7}		
	0.016 179 388 247	$1.439 2 \times 10^{-7}$		
	0.016 177 6	1.3×10^{-7}	4.50	0.830 1
	0.016 179 7	1.46×10^{-7}	4.446	0.830 471
	0.016 179 27	1.426×10^{-7}	4.456 1	0.830 463 4
	0.016 145	1.45×10^{-7}	4.451	0.829 5
	0.016 178 0	1.47×10^{-7}	4.443	0.830 44
	0.016 179 14	1.427×10^{-7}	4.455 9	0.830 461
$2.139 3 \times 10^{-4}$ au (1.1×10^6 V/cm)	0.015 842 9	1.09×10^{-6}		0.8339
	0.015 860 468	1.057×10^{-6}		
	—	—		
	0.015 857	0.97×10^{-6}	3.48	0.833 4
	0.015 861 0	1.07×10^{-6}	3.428	0.833 84
	0.015 860 29	1.048×10^{-6}	3.438 7	0.833 826
	0.015 815	1.10×10^{-6}	3.422	0.832 7
	0.015 857	1.08×10^{-6}	3.423	0.833 78
	0.015 859 7	1.051×0^{-6}	3.437 9	0.833 816
$2.528 2 \times 10^{-4}$ au (1.3×10^6 V/cm)	0.015 255	1.78×10^{-5}		0.841 05
	0.015 269 204	$1.756 0 \times 10^{-5}$		
	—	—		
	0.015 260	1.6×10^{-5}	2.01	0.840 4
	0.015 270 2	1.79×10^{-5}	1.96	0.840 99
	0.015 268 8	1.748×10^{-5}	1.975 2	0.840 970
	0.015 1	2.0×10^{-5}	1.92	0.839 2
	0.015 258	1.85×10^{-5}	1.95	0.840 79
	0.015 268 2	1.739×10^{-5}	1.974 5	0.840 961

(Continued)

Table 8.4. (*Continued*)

F	$-E$	Γ	K	Z_1
$2.917\,2 \times 10^{-4}$ au (1.5×10^{6} V/cm)	0.014 735	0.995×10^{-4}		0.848 86
	0.014 740 243	$0.976\,51 \times 10^{-4}$		
	—	—		
	0.014 725	0.94×10^{-4}	1.05	0.848 0
	0.014 742	1.02×10^{-4}	1.01	0.848 72
	0.014 739 6	0.999×10^{-4}	1.021	0.848 683
	0.014 62	1.2×10^{-4}	0.93	0.846 2
	0.014 733	0.993×10^{-4}	1.00	0.848 58
	0.014 78	0.8×10^{-4}	1.06	0.849 3
$3.306\,1 \times 10^{-4}$ au (1.7×10^{6} V/cm)	0.014 27	3.03×10^{-4}		0.857 24
	0.014 242 49	$2.785\,3 \times 10^{-4}$		
	—	—		
	0.014 22	2.9×10^{-4}	0.40	0.855 6
	0.014 246	3.15×10^{-4}	0.355	0.856 52
	0.014 242 8	3.08×10^{-4}	0.367	0.856 471
	0.014 12	3.06×10^{-4}	0.29	0.854 0
	0.014 32	1.9×10^{-4}	0.43	0.857 7
	0.014 5	1.0×10^{-4}	0.6	0.860

Korsch and Möhlenkamp (1983) have used a semiclassical treatment of complex energy states to calculate results corresponding to those in our Table 8.4. The results in their Table 1 are in satisfactory agreement with the first-order approximation of our phase-integral results.

Table 8.5 gives results concerning the Stark effect of the hydrogen atom for four states with the principal quantum number $n = |m|+1+n_1+n_2 = 25$, for one state with $n = 12$, and for one state with $n = 7$. For each state and each value of the electric field strength F the first line gives the numerical results obtained by Damburg and Kolosov (1981); see their Fig. 1 for the state $m = 0, n_1 = 0, n_2 = 24$; see their Fig. 2 for the state $|m| = 1, n_1 = 11$, $n_2 = 12$; see their Fig. 3 for the state $|m| = 1, n_1 = 12, n_2 = 11$, see their Fig. 4 and their Table 1 for the state $m = 0, n_1 = 24, n_2 = 0$, see their Table 2 for the state $|m| = 2, n_1 = 8, n_2 = 1$, see their Fig. 5 and their table 3 for the state $|m| = 1, n_1 = 3, n_2 = 2$. The lines 2–7 give our phase-integral

Table 8.5. $n = 25$ (four states), $n = 12$ (one state) and $n = 7$ (one state).

$\|m\|$	n_1	n_2	$F \times 10^6$	$-E$	Γ	K	Z_1
0	0	24	0.36	$1.175\,31 \times 10^{-3}$	0.710×10^{-5}		
				$1.175\,313 \times 10^{-3}$	1.579×10^{-5}	$-1.502\,6$	0.024 299
				$1.175\,310\,530 \times 10^{-3}$	$1.580\,5 \times 10^{-5}$	$-1.503\,8$	0.024 317 640 5
				$1.175\,310\,527\,5 \times 10^{-3}$	$1.580\,5 \times 10^{-5}$	$-1.503\,8$	0.024 317 641 49
				$1.175\,8 \times 10^{-3}$	1.54×10^{-5}	-1.43	0.024 304
				$1.175\,36 \times 10^{-3}$	1.57×10^{-5}	-1.496	0.024 318 1
				$1.175\,24 \times 10^{-3}$	1.55×10^{-5}	-1.515	0.024 316 9
1	11	12	0.46	$8.904\,24 \times 10^{-4}$	$0.856\,7 \times 10^{-5}$		
				$8.904\,37 \times 10^{-4}$	1.989×10^{-5}	-1.601	0.553 71
				$8.904\,449\,1 \times 10^{-4}$	$1.991\,8 \times 10^{-5}$	$-1.602\,7$	0.553 728 385
				$8.904\,448\,98 \times 10^{-4}$	$1.991\,8 \times 10^{-5}$	$-1.602\,7$	0.553 728 383 5
				8.910×10^{-4}	1.95×10^{-5}	-1.53	0.553 83
				$8.905\,1 \times 10^{-4}$	1.98×10^{-5}	-1.595	0.553 74
				$8.903\,8 \times 10^{-4}$	1.96×10^{-5}	-1.610	0.553 71
1	12	11	0.46	$8.538\,66 \times 10^{-4}$	$0.635\,8 \times 10^{-5}$		
				$8.538\,63 \times 10^{-4}$	1.138×10^{-5}	-1.036	0.593 97
				$8.538\,747\,1 \times 10^{-4}$	$1.140\,2 \times 10^{-5}$	$-1.037\,7$	0.593 992 161
				$8.538\,746\,94 \times 10^{-4}$	$1.140\,2 \times 10^{-5}$	$-1.037\,7$	0.593 992 158 7
				8.545×10^{-4}	1.09×10^{-5}	-0.96	0.594 1
				$8.538\,5 \times 10^{-4}$	1.10×10^{-5}	-1.041	0.593 986
				8.535×10^{-4}	1.05×10^{-5}	-1.08	0.593 90

(Continued)

Table 8.5. *(Continued)*

$\|m\|$	n_1	n_2	$F \times 10^6$	$-E$	Γ	K	Z_1
0	24	0	0.6	$3.156\,411\,08 \times 10^{-4}$	9.48×10^{-11}		
				$3.155\,6 \times 10^{-4}$	8.3×10^{-11}	5.69	0.987 90
				$3.156\,44 \times 10^{-4}$	9.71×10^{-11}	5.614	0.987 965
				$3.156\,398 \times 10^{-4}$	$9.314\,4 \times 10^{-11}$	5.634 6	0.987 963 1
				3.154×10^{-4}	8.7×10^{-11}	5.67	0.987 82
				$3.156\,36 \times 10^{-4}$	9.73×10^{-11}	5.613	0.987 961
				$3.156\,406 \times 10^{-4}$	$9.314\,3 \times 10^{-11}$	5.634 7	0.987 963 5
2	8	1	7.779 074 3	$2.657\,073\,8 \times 10^{-3}$	2.15×10^{-8}		
				$2.656\,8 \times 10^{-3}$	2.08×10^{-8}	4.03	0.833 63
				$2.657\,078 \times 10^{-3}$	2.150×10^{-8}	4.018 3	0.833 732 4
				$2.657\,073\,7 \times 10^{-3}$	2.148×10^{-8}	4.018 7	0.833 732 06
				2.655×10^{-3}	2.26×10^{-8}	4.00	0.833 4
				$2.656\,96 \times 10^{-3}$	2.16×10^{-8}	4.016	0.833 72
				$2.657\,05 \times 10^{-3}$	2.151×10^{-8}	4.018 3	0.833 730
1	3	2	6.5	$1.014\,055 \times 10^{-2}$	3.597×10^{-5}		
				$1.013\,9 \times 10^{-2}$	3.68×10^{-5}	0.94	0.632 7
				$1.014\,058 \times 10^{-2}$	$3.741\,4 \times 10^{-5}$	0.927 55	0.632 960 4
				$1.014\,056\,0 \times 10^{-2}$	$3.740\,9 \times 10^{-5}$	0.927 63	0.632 959 93
				1.010×10^{-2}	4.6×10^{-5}	0.82	0.631 9
				$1.014\,09 \times 10^{-2}$	3.5×10^{-5}	0.929	0.632 968
				1.016×10^{-2}	2.7×10^{-5}	0.99	0.633 4

results both with $\tilde{\phi}$ included (the lines 2, 3 and 4 giving the results in the first-, third- and fifth-order approximations, respectively) and with $\tilde{\phi}$ neglected (the lines 5, 6 and 7 giving the results in the first-, third- and fifth-order approximations, respectively).

For the state $m = 0, n_1 = 0, n_2 = 24$ the value of E given by Damburg and Kolosov (1981) is somewhat more accurate than our optimum value when $\tilde{\phi}$ is neglected. In the third- and fifth-order approximations with $\tilde{\phi}$ included we get probably three more reliable digits than Damburg and Kolosov.

For the states $|m| = 1, n_1 = 11, n_2 = 12$ and $|m| = 1, n_1 = 12, n_2 = 11$ the last two digits in the values of E obtained by Damburg and Kolosov (1981) differ from the results obtained by us in the third- and fifth-order approximations with $\tilde{\phi}$ included. This indicates that for these states the values of E given by Damburg and Kolosov (1981) may not be quite correct.

For the state $m = 0, n_1 = 24, n_2 = 0$ the value of E obtained by Damburg and Kolosov (1981) agrees best with the phase-integral value of the fifth-order approximation with $\tilde{\phi}$ neglected. This is strange, since $\tilde{\phi}$ should be important also when the barrier is thick, if one wants results of great accuracy.

For the states $|m| = 2$, $n_1 = 8$, $n_2 = 1$ and $|m| = 1$, $n_1 = 3$, $n_2 = 2$ the fifth-order (with $\tilde{\phi}$ included) phase-integral values for E agree satisfactorily with the corresponding values obtained by Damburg and Kolosov (1981).

Tables 8.6(a), (b) give results concerning the Stark effect of the hydrogen atom for states with the principal quantum number $n = |m|+1+n_1+n_2 = 10$ and $n = 25$ (Table 8.6(a)) and with $n = 13, n = 14$ and $n = 30$ (Table 8.6(b)). The field strength F is expressed in au in Table 8.6(a) but originally in V/cm in Table 8.6(b). For each state and each value of the electric field strength F the first line gives the "theoretical" results obtained by Kolosov (1983); when he gives two values of Γ, obtained from his Eqs. (31) and (32), we quote only that obtained from the more accurate Eq. (31). The second line quotes the numerically exact results given by Kolosov (1983). The lines 3–8 give our phase-integral results both with $\tilde{\phi}$ included (the lines 3, 4 and 5 giving the results in the first-, third- and fifth-order approximations, respectively) and with $\tilde{\phi}$ neglected (the lines

Table 8.6(a). $an = 5, n = 10$ and $n = 25$.

$\lvert m\rvert$	n	n_1	n_2	F	$-E \times 10^2$	Γ	K	Z_1
0	10	0	9	1.0×10^{-5}au	0.648 2	2.54×10^{-8}		
					0.647 541	2.40×10^{-8}		
					0.647 56	2.38×10^{-8}	4.160	0.057 19
					0.647 541 47	$2.404\,5 \times 10^{-8}$	4.155 2	0.057 281 20
					0.647 541 434	$2.404\,5 \times 10^{-8}$	4.155 2	0.057 281 224
					0.647 2	2.6×10^{-8}	4.12	0.057 17
					0.647 528	2.416×10^{-8}	4.153	0.057 280 6
					0.647 539	2.407×10^{-8}	4.154 8	0.057 281 1
0	10	0	9	1.1×10^{-5}au	0.664 8	7.46×10^{-7}		
					0.664 598	7.17×10^{-7}		
					0.664 62	7.13×10^{-7}	2.397	0.057 95
					0.664 597 57	$7.188\,3 \times 10^{-7}$	2.392 9	0.058 052 66
					0.664 597 526	$7.188\,3 \times 10^{-7}$	2.392 9	0.058 052 688
					0.664 1	8.3×10^{-7}	2.33	0.057 93
					0.664 54	7.3×10^{-7}	2.386	0.058 050
					0.664 58	7.21×10^{-7}	2.391 5	0.058 052 2
0	25	0	24	3×10^{-7}au	0.110 00	1.30×10^{-8}		
					0.109 986 17	1.21×10^{-8}		
					0.109 986 6	1.204×10^{-8}	3.001	0.023 502
					0.109 986 172 3	$1.207\,1 \times 10^{-8}$	2.999 7	0.023 518 354 5
					0.109 986 172 16	$1.207\,1 \times 10^{-8}$	2.999 7	0.023 518 355 23
					0.109 96	1.3×10^{-8}	2.96	0.023 499
					0.109 984	1.219×10^{-8}	2.996	0.023 518 2
					0.109 985 7	1.210×10^{-8}	2.998 8	0.023 518 30

Table 8.6(b). $n = 13, n = 14$ and $n = 30$.

$\lvert m \rvert$	n	n_1	n_2	F	$-E \times 10^3$	Γ	K	Z_1
2	13	4	6	22 000 V/cm $\approx 4.278 \times 10^{-6}$au	3.219	6.13×10^{-9}		
					3.232 81	5.48×10^{-9}		
					3.232 83	5.42×10^{-9}	4.536	0.468 88
					3.232 809 15	$5.472 0 \times 10^{-9}$	4.531 2	0.468 938 774
					3.232 808 968	$5.471 9 \times 10^{-9}$	4.531 2	0.468 938 770 0
					3.231 4	5.8×10^{-9}	4.50	0.468 79
					3.232 76	5.49×10^{-9}	4.530 0	0.468 936
					3.232 800	5.476×10^{-9}	4.530 9	0.468 938 2
				25 000 V/cm $\approx 4.862 \times 10^{-6}$au	3.280	8.18×10^{-7}		
					3.294 88	7.74×10^{-7}		
					3.294 86	7.72×10^{-7}	1.963	0.475 88
					3.294 877 8	$7.779 1 \times 10^{-7}$	1.959 0	0.475 943 098
					3.294 877 56	$7.779 1 \times 10^{-7}$	1.959 0	0.475 943 091 3
					3.292	9.1×10^{-7}	1.89	0.475 7
					3.294 5	8.0×10^{-7}	1.950	0.475 92
					3.294 869	7.73×10^{-7}	1.958 8	0.475 942 6
2	14	1	10	15 000 V/cm $\approx 2.917 \times 10^{-6}$au	3.190	9.66×10^{-9}		
					3.184 70	8.91×10^{-9}		
					3.184 75	8.85×10^{-9}	4.133	0.203 44
					3.184 705 02	$8.914 3 \times 10^{-9}$	4.129 4	0.203 488 171
					3.184 704 955	$8.914 3 \times 10^{-9}$	4.129 4	0.203 488 174 8
					3.183 6	9.6×10^{-9}	4.10	0.203 40

(*Continued*)

Table 8.6(b). *(Continued)*

$\|m\|$	n	n_1	n_2	F	$-E \times 10^3$	Γ	K	Z_1
					3.184 66	8.95×10^{-9}	4.128	0.203 486 7
					3.184 695	8.923×10^{-9}	4.129 1	0.203 487 9
				16 000 V/cm $\approx 3.111 \times 10^{-6}$au	3.241	1.64×10^{-7}		
					3.236 40	1.55×10^{-7}		
					3.236 43	1.545×10^{-7}	2.654	0.205 24
					3.236 397 94	$1.554\,7 \times 10^{-7}$	2.650 9	0.205 291 582
					3.236 397 865	$1.554\,7 \times 10^{-7}$	2.650 9	0.205 291 586 5
					3.234 7	1.7×10^{-7}	2.60	0.205 19
					3.236 23	1.58×10^{-7}	2.646	0.205 286
					3.236 36	1.560×10^{-7}	2.649 6	0.205 290
				17 000 V/cm $\approx 3.306 \times 10^{-6}$au	3.295	1.41×10^{-6}		
					3.290 94	1.35×10^{-6}		
					3.290 96	1.361×10^{-6}	1.489	0.207 11
					3.290 942 22	$1.367\,7 \times 10^{-6}$	1.486 3	0.207 160 116
					3.290 942 144	$1.367\,7 \times 10^{-6}$	1.486 3	0.207 160 121 3
					3.288	1.7×10^{-6}	1.41	0.207 0
					3.290 6	1.39×10^{-6}	1.476	0.207 15
					3.291 2	1.30×10^{-6}	1.494	0.207 168
0	30	0	29	720 V/cm $\approx 1.4 \times 10^{-7}$au	0.757 3	1.57×10^{-10}		
					0.757 161 6	1.54×10^{-10}		
					0.757 164	1.540×10^{-10}	4.943	0.019 492

(Continued)

Table 8.6(b). (*Continued*)

$\|m\|$	n	n_1	n_2	F	$-E \times 10^2$	Γ	K	Z_1
					0.757 161 593 9	1.5445×10^{-10}	4.941 1	0.019 503 138 9
					0.757 161 593 4	1.5445×10^{-10}	4.941 1	0.019 503 139 2
					0.757 08	1.63×10^{-10}	4.92	0.019 491
					0.757 159	1.547×10^{-10}	4.940 4	0.019 503 11
					0.757 161 2	1.5450×10^{-10}	4.941 0	0.019 503 135
				740 V/cm $\approx 1.439 \times 10^{-7}$au	0.763 82	1.60×10^{-9}		
					0.763 737 1	1.57×10^{-9}		
					0.763 739	1.571×10^{-9}	3.753	0.019 577
					0.763 737 122 9	1.5750×10^{-9}	3.751 4	0.019 588 310 4
					0.763 737 122 3	1.5750×10^{-9}	3.751 4	0.019 588 310 7
					0.763 63	1.69×10^{-9}	3.72	0.019 575
					0.763 732	1.582×10^{-9}	3.749 7	0.019 588 24
					0.763 735 8	1.577×10^{-9}	3.751 0	0.019 588 29
				760 V/cm $\approx 1.478 \times 10^{-7}$au	0.770 49	1.23×10^{-8}		
					0.770 453 1	1.21×10^{-8}		
					0.770 455	1.210×10^{-8}	2.695	0.019 663
					0.770 453 146 6	1.2130×10^{-8}	2.693 8	0.019 674 885 9
					0.770 453 146 0	1.2130×10^{-8}	2.693 8	0.019 674 886 3
					0.770 31	1.34×10^{-8}	2.65	0.019 661
					0.770 439	1.227×10^{-8}	2.690	0.019 674 71
					0.770 450	1.216×10^{-8}	2.692 7	0.019 674 84

(*Continued*)

Table 8.6(b). *(Continued)*

$\|m\|$	n	n_1	n_2	F	$-E \times 10^2$	Γ	K	Z_1
				780 V/cm $\approx 1.517 \times 10^{-7}$au	0.777 35	6.98×10^{-8}		
					0.777 349 7	6.87×10^{-8}		
					0.777 351 0	6.91×10^{-8}	1.771	0.019 751
					0.777 349 712 9	$6.922\,5 \times 10^{-8}$	1.769 8	0.019 763 352 3
					0.777 349 712 3	$6.922\,5 \times 10^{-8}$	1.769 8	0.019 763 352 7
					0.777 1	8.0×10^{-8}	1.71	0.019 749
					0.777 32	7.1×10^{-8}	1.762	0.019 763 0
					0.777 355	6.83×10^{-8}	1.771	0.019 763 41

6, 7 and 8 giving the results in the first-, third- and fifth- order approximations, respectively).

For the state $m = 0, n_1 = 0, n_2 = 9$, i.e., $n = 10$, with the electric field strength $F = 1.0 \times 10^{-5}$ au in Table 1 in Kolosov (1983) there is in the value of E_{numer} probably a misprint which we have corrected in our Table 8.6(a).

In Table 8.6(a) the electric field strengths F (expressed in au) are simple numbers, which we have considered as exact values in atomic units. In Table 8.6(b) the electric field strengths F are originally given in V/cm and have been converted into au by conversion factors which we do not know exactly. After certain difficulties with these conversion factors we have in the third- and fifth-order approximations with $\tilde{\phi}$ included been able to reproduce all digits in the numerically exact values of E obtained by Kolosov (1983). To achieve this result, we have used the conversion factor 1 au $= 5.142 \times 10^{-9}$ V/cm for the states with $n = 13$ and $n = 14$ and the conversion factor 1 au $= 5.142\,260\,3 \times 10^{-9}$ V/cm for the state with $n = 30$; it is not possible to reproduce all digits in the numerically exact results by the use of one and the same conversion factor.

Table 8.7 gives results concerning the Stark effect of the hydrogen atom for one state with the principal quantum number $n = |m| + 1 + n_1 + n_2 = 25$ and the field strength $F = 2514$ V/cm $\approx 4.889 \times 10^{-7}$ au and one state with $n = |m| + 1 + n_1 + n_2 = 30$ and $F = 800$ V/cm $\approx 1.556 \times 10^{-7}$ au. To convert these field strengths from V/cm into au we have used the conversion factor 1 au $= 5.142\,260\,3 \times 10^{-9}$ V/cm (obtained as private communication from Professor Damburg). For each state and each electric field strength F the first line in our Table 8.7 gives the value of E obtained by combination of Rayleigh–Schrödinger perturbation theory of the order 24 with Padé approximant technique (Silverstone and Koch 1979), the second line gives the numerical value of E calculated according to the method of Damburg and Kolosov (1976a) and obtained as private communication from Damburg to Nanny Fröman in a letter of 22 February 1985, and the lines 3–8 give our phase-integral results both with $\tilde{\phi}$ included (the lines 3, 4 and 5 giving the results in the first-, third- and fifth-order approximations, respectively) and with $\tilde{\phi}$ neglected (the lines 6, 7

Table 8.7. $n = 25$ and $n = 30$.

$\lvert m\rvert$	n	n_1	n_2	F	$-E \times 10^4$	Γ	K	Z_1
1	25	21	2	2514 V/cm $\approx 4.889 \times 10^{-7}$au	4.987 03			
					—			
					4.986 8	5.34×10^{-11}	5.91	0.913 11
					4.987 038 4	$5.464\,3 \times 10^{-11}$	5.895 0	0.913 140 05
					4.987 037 07	$5.463\,1 \times 10^{-11}$	5.895 1	0.913 139 992
					4.985 1	5.60×10^{-11}	5.88	0.913 03
					4.987 00	5.471×10^{-11}	5.894 5	0.913 138
					4.987 034	$5.463\,9 \times 10^{-11}$	5.895 1	0.913 139 8
0	30	0	29	800 V/cm $\approx 1.556 \times 10^{-7}$au	7.844 68			
					7.844 648 04			
					7.844 656	2.849×10^{-7}	0.981	0.019 842
					7.844 648 053	$2.853\,2 \times 10^{-7}$	0.980 29	0.019 854 161 3
					7.844 648 046	$2.853\,2 \times 10^{-7}$	0.980 29	0.019 854 161 7
					7.842	3.4×10^{-7}	0.90	0.019 838
					7.844 651	2.73×10^{-7}	0.980 36	0.019 854 165
					7.846 1	2.3×10^{-7}	1.02	0.019 856 0

and 8 giving the results in the first-, third- and fifth-order approximations, respectively).

From Figs 1, 2 and table 1 in Silverstone and Koch (1979) it is seen that merely Rayleigh–Schrödinger perturbation theory, even of very high orders (up to 24), cannot give values of E that are sufficiently accurate to reproduce the experimental results. Therefore we have not included values of E calculated in that way in our Table 8.7. The values obtained from the combination of Rayleigh–Schrödinger perturbation theory of the order 24 with Padé approximant technique in the first line associated with each state and each field strength in our Table 8.7 are much more accurate. Still more accurate and more reliable than the values obtained by the combination of Rayleigh–Schrödinger perturbation theory of large orders with Padé approximant technique, quoted in our Table 8.7, are the results that can be obtained by the numerical method of Damburg and Kolosov (1976a) or by the phase-integral method. For the state with $n = 30$ the digits in our Table 8.7 indicate that in the third- and fifth-order approximations with $\tilde{\phi}$ included, the phase-integral value of E reproduces the numerical value calculated by Damburg and Kolosov (1976a).

For the state $n = 30, m = 0, n_1 = 0, n_2 = 29$ in the electric field $F = 800$ V/cm $\approx 1.556 \times 10^{-7}$ au, Silverstone and Koch (1979) quote (in the caption to their Fig. 2) the energy value $E = -7.844\,648 \times 10^{-4}$ au, which has been calculated by Damburg and Kolosov (private communication from Damburg to Silverstone and Koch). Silverstone and Koch (1979) say that this energy value was obtained by the phase shift method, but Damburg has informed us (private communication to Nanny Fröman in a letter of 26 May 1980) that the energy value in question was determined as the value of E for which the quantity B in Eq. (11) in Damburg and Kolosov (1980) is minimum. In a letter to Nanny Fröman of 22 February 1985 Damburg communicated the improved value quoted in Table 8.7. We have calculated the positions of the Stark levels by means of the accurate formula (5.44) along with (5.42). The state $m = 0, n_1 = 0, n_2 = 29, n = |m| + 1 + n_1 + n_2 = 30$ is also discussed in Chapter 1 of this book.

Table 8.8 gives results concerning the Stark effect for four states of the hydrogen atom with the quantum numbers $m = 0, n_1 = 7$ and $n_2 = 1, 2, 3$ and 4, i.e., $n = |m| + 1 + n_1 + n_2 = 9$, 10, 11 and 12 in

Table 8.8. $F = 1.5 \times 10^{-5}$au, $m = 0, n_1 = 7, n_2 = 1, 2, 3, 4, n = 9, 10, 11, 12.$

n	$-E \times 10^3$	Γ	K	Z_1
9	5.065 779 842 367 91	3.7×10^{-16}		
	5.065 84	3.5×10^{-16}	13.52	0.854 5
	5.065 784	3.694×10^{-16}	13.496	0.854 610 0
	5.065 779 63	$3.688 3 \times 10^{-16}$	13.497	0.854 609 765
	5.064	3.64×10^{-16}	13.50	0.854 4
	5.065 771	3.695×10^{-16}	13.496	0.854 609 2
	5.065 779 69	$3.688 3 \times 10^{-16}$	13.497	0.854 609 768
10	4.097 121 64	6.368×10^{-8}		
	4.096 8	6.2×10^{-8}	3.76	0.794 87
	4.097 126	6.371×10^{-8}	3.750 0	0.794 996 8
	4.097 121 5	6.369×10^{-8}	3.750 1	0.794 996 56
	4.093	6.8×10^{-8}	3.72	0.794 6
	4.096 9	6.41×10^{-8}	3.748	0.794 98
	4.097 07	6.38×10^{-8}	3.749 6	0.794 993
11	3.600 7	1.353×10^{-4}		
	3.600 8	1.89×10^{-4}	−1.083	0.763 0
	3.601 436	$1.908 1 \times 10^{-4}$	−1.089 9	0.763 190 2
	3.601 431 4	$1.908 0 \times 10^{-4}$	−1.089 8	0.763 189 94
	3.610	1.80×10^{-4}	−0.99	0.763 6
	3.601 1	1.85×10^{-4}	−1.093	0.763 17
	3.596	1.77×10^{-4}	−1.14	0.762 9
12	3.100	1.637×10^{-3}		
	2.905	1.70×10^{-2}	−7.72	0.716 71
	2.903 689	$1.721 5 \times 10^{-2}$	−7.739 9	0.716 771 3
	2.903 693 1	$1.721 4 \times 10^{-2}$	−7.739 8	0.716 771 63
	2.908	1.68×10^{-2}	−7.69	0.716 9
	2.903 72	$1.721 4 \times 10^{-2}$	−7.739 6	0.716 773
	2.903 695	$1.721 4 \times 10^{-2}$	−7.739 8	0.716 771 8

the electric field $F = 1.5 \times 10^{-5}$ au. For each state the first line gives results calculated by Luc-Koenig and Bachelier (1980a) and given in the caption to their Fig. 3; their notation E_0 is defined in their Eq. (1). The lines 2–7 give our phase-integral results both with $\tilde{\phi}$ included (the lines 2, 3 and 4 giving the results in the first-, third- and fifth-order approximations, respectively) and with $\tilde{\phi}$ neglected (the lines 5, 6 and 7 giving the results in the first-, third- and fifth-order approximations, respectively).

For the state with $n_2 = 1, \text{i.e.}, n = 9$, the barrier is very thick (K positive and large), and the phase-integral value of E that agrees best with the value obtained by Luc-Koenig and Bachelier is obtained in the fifth-order approximation with $\tilde{\phi}$ neglected. For the state $n_2 = 2$, i.e., $n = 10$, the phase-integral results in the fifth-order approximation with $\tilde{\phi}$ retained agree satisfactorily with the results obtained by Luc-Koenig and Bachelier. For the states with $n_2 = 3$ and $n_2 = 4$, i.e., $n = 11$ and $n = 12$, respectively, our optimum phase-integral results with $\tilde{\phi}$ included seem to be more accurate than the values obtained by Luc-Koenig and Bachelier. For the state with $n_2 = 4$, i.e., $n = 12$, the value $\Gamma = 1.637 \times 10^{-3}$ given by Luc-Koenig and Bachelier differs essentially from our value. The reason is that the definition of the half-width used by Luc-Koenig and Bachelier is not correct for this resonance, as Luc-Koenig told us in a letter of 29 August 2004. Nor is our formula (5.53) in Chapter 5 for the half-width useful, since the barrier is underdense.

Tables 8.9(a), (b) give results concerning the Stark effect for the two states $m = 0, n_1 = 9, n_2 = 0$ (Table 8.9(a)) and $m = 0, n_1 = 0, n_2 = 9$ (Table 8.9(b)) of the hydrogen atom with the principal quantum number $n = |m| + 1 + n_1 + n_2 = 10$ in electric fields of various strengths. For each state and each value of the field strength F, the first line gives the numerical results obtained by Damburg and Kolosov (1976b) and quoted by Luc-Koenig and Bachelier (1980a) in their Table 1. The second line gives the numerical results calculated by Luc-Koenig and Bachelier (1980a) and given in their Table 1. The lines 3–8 give our phase-integral results both with $\tilde{\phi}$ included (the lines 3, 4 and 5 giving the results in the first-, third- and fifth-order approximations, respectively) and with $\tilde{\phi}$ neglected (the lines 6, 7 and 8 giving the results in the first-, third- and fifth-order approximations, respectively).

The value of Γ due to Luc-Koenig and Bachelier (1980a) for the field strength $F = 2.0 \times 10^{-5}$ au is approximately ten times larger than the corresponding value of Γ obtained by Damburg and Kolosov (1976b) and by us. One may therefore suspect that there is a misprint amounting to a factor of ten in the value given by Luc-Koenig and Bachelier (1980a) for the particular value of Γ in question. This has been confirmed by Luc-Koenig in a letter to us of 29 August 2004, and therefore we give in our Table 8.9(a) the revised value. The values

Table 8.9(a). $m = 0, n_1 = 9, n_2 = 0, n = 10.$

$F \times 10^5$	$-E \times 10^3$	Γ	K	Z_1
2.0	2.585 573 8	1.900×10^{-7}		
	2.585 573 9	1.903×10^{-7}		
	2.584	1.7×10^{-7}	3.29	0.966 1
	2.585 7	1.95×10^{-7}	3.211	0.966 34
	2.585 52	1.867×10^{-7}	3.232	0.966 324
	2.579	1.87×10^{-7}	3.24	0.965 7
	2.585 1	1.97×10^{-7}	3.205	0.966 29
	2.585 46	1.871×10^{-7}	3.231	0.966 318
2.5	2.064 665	1.250×10^{-5}		
	2.064 666	1.293×10^{-5}		
	2.061	1.2×10^{-5}	1.02	0.971 7
	2.065 1	1.33×10^{-5}	0.95	0.972 17
	2.064 5	1.27×10^{-5}	0.97	0.972 12
	2.05	1.5×10^{-5}	0.91	0.970 8
	2.063 8	1.29×10^{-5}	0.94	0.972 07
	2.070	1.0×10^{-5}	1.01	0.972 5
3.0	1.563 15	6.956×10^{-5}		
	1.563 17	9.010×10^{-5}		
	1.558	8.3×10^{-5}	−0.18	0.977 4
	1.563 6	9.4×10^{-5}	−0.26	0.978 03
	1.563 22	8.9×10^{-5}	−0.23	0.978 003
	1.564 0	6.1×10^{-5}	−0.13	0.977 9
	1.54	6.3×10^{-5}	−0.4	0.976
	1.51	4.9×10^{-5}	−0.6	0.974
3.5	1.040 77	1.571×10^{-4}		
	1.041	2.388×10^{-4}		
	1.037	2.7×10^{-4}	−1.02	0.980 7
	1.039 6	3.0×10^{-4}	−1.13	0.981 1
	1.043	2.9×10^{-4}	−1.08	0.981 4
	1.05	2.5×10^{-4}	−0.92	0.981 7
	1.038	2.9×10^{-4}	−1.14	0.981 0
	1.033	2.7×10^{-4}	−1.15	0.980 6
4.0	0.480 07	2.607×10^{-4}		
	0.481	5.45×10^{-4}		
	0.481 9	6.5×10^{-4}	−1.78	0.980 8
	0.476	7.25×10^{-4}	−1.93	0.980 59
	0.483	7.30×10^{-4}	−1.89	0.981 05
	0.50	6.3×10^{-4}	−1.69	0.981 7
	0.477	7.22×10^{-4}	−1.92	0.980 66
	0.481 8	7.23×10^{-4}	−1.90	0.980 97

Table 8.9(b). $m = 0, n_1 = 0, n_2 = 9, n = 10.$

$F \times 10^5$	$-E \times 10^3$	Γ	K	Z_1
1.2	6.826 384 6	7.145×10^{-6}		
	6.826 385	7.327×10^{-6}		
	6.826 5	7.29×10^{-6}	1.129	0.058 75
	6.826 385 0	$7.336\,9 \times 10^{-6}$	1.125 3	0.058 854 80
	6.826 384 56	$7.336\,9 \times 10^{-6}$	1.125 3	0.058 854 828
	6.817	9.1×10^{-6}	1.03	0.058 71
	6.825 8	7.2×10^{-6}	1.119	0.058 852
	6.829	6.2×10^{-6}	1.16	0.058 867
1.4	7.202 314	6.797×10^{-5}		
	7.202	9.861×10^{-5}		
	7.202 38	9.79×10^{-5}	−0.492	0.060 37
	7.202 315 6	$9.834\,5 \times 10^{-5}$	−0.494 77	0.060 487 05
	7.202 315 05	$9.834\,5 \times 10^{-5}$	−0.494 77	0.060 487 081
	7.21	8.34×10^{-5}	−0.41	0.060 41
	7.19	8.32×10^{-5}	−0.57	0.060 46
	7.18	6.9×10^{-5}	−0.71	0.060 40
1.6	7.535 1	1.795×10^{-4}		
	7.536	3.4×10^{-4}		
	7.535 3	4.75×10^{-4}	−1.837	0.061 77
	7.535 064	$4.766\,8 \times 10^{-4}$	−1.841 3	0.061 901 12
	7.535 063 4	$4.766\,8 \times 10^{-4}$	−1.841 3	0.061 901 161
	7.544	4.6×10^{-4}	−1.76	0.061 81
	7.536 3	4.76×10^{-4}	−1.83	0.061 906
	7.534 5	4.73×10^{-4}	−1.85	0.061 899
1.8	7.769 7	$0.312\,9 \times 10^{-3}$		
	7.770	0.79×10^{-3}		
	7.770 2	1.81×10^{-3}	−3.480	0.062 76
	7.769 723 9	$1.822\,4 \times 10^{-3}$	−3.485 8	0.062 896 19
	7.769 723 04	$1.822\,4 \times 10^{-3}$	−3.485 8	0.062 896 242
	7.777	1.78×10^{-3}	−3.42	0.062 79
	7.770 2	$1.822\,4 \times 10^{-3}$	−3.482	0.062 898
	7.769 9	$1.822\,5 \times 10^{-3}$	−3.485	0.062 896 8
2.0	7.887	$0.466\,0 \times 10^{-3}$		
	7.981	2.7×10^{-3}		
	7.888	7.39×10^{-3}	−5.52	0.063 27
	7.887 121	$7.432\,3 \times 10^{-3}$	−5.531 1	0.063 417 92
	7.887 119 9	$7.432\,3 \times 10^{-3}$	−5.531 1	0.063 417 987
	7.894	7.2×10^{-3}	−5.48	0.063 29
	7.887 3	7.431×10^{-3}	−5.530	0.063 418 5
	7.887 14	$7.432\,3 \times 10^{-3}$	−5.530 9	0.063 418 08

of Γ obtained by Luc-Koenig and Bachelier (1980a) and quoted in our Table 8.9(a) are for the field strengths $2.0 \times 10^{-5}, 2.5 \times 10^{-5}$ and 3.0×10^{-5} the real half-widths and for the field strengths 3.5×10^{-5} and 4.0×10^{-5} the Lorentzian half-widths. The values of Γ obtained from Luc-Koenig and Bachelier (1980a) and quoted in our Table 8.9(b) are for the field strengths 1.2×10^{-5} and 1.4×10^{-5} the real half-widths and for the field strengths $1.6 \times 10^{-5}, 1.8 \times 10^{-5}$ and 2.0×10^{-5} the Lorentzian half-widths.

Tables 8.10(a–l) give results concerning the Stark effect of the hydrogen atom when $|m| = 1$ for the states with $n = 14$ (Table 8.10(a)), $n = 13$ (Table 8.10(b)) and $n = 12$ (Table 8.10(c)) in the electric field $F = 0.5 \times 10^{-5}$ au, for the states with $n = 12$ (Table 8.10(d)), $n = 11$ (Table 8.10(e)) and $n = 10$ (Table 8.10(f)) in the electric field $F = 1.0 \times 10^{-5}$ au, for the states with $n = 11$ (Table 8.10(g)), $n = 10$ (Table 8.10(h)) and $n = 9$ (Table 8.10(i)) in the electric field $F = 1.5 \times 10^{-5}$ au, and for the states with $n = 10$ (Table 8.10(j)), $n = 9$ (Table 8.10(k)) and $n = 8$ (Table 8.10(l)) in the electric field $F = 2.0 \times 10^{-5}$ au. For each state and each value of the field strength F, the first line in all these tables gives the results obtained by Luc-Koenig and Bachelier (1980a) and presented in their Table 5. In our Tables 8.10(a–f) and 8.10(j–l) the lines 2–7 give our phase-integral results both with $\tilde{\phi}$ included (the lines 2, 3, 4 giving the results in the first-, third- and fifth-order approximations, respectively), and with $\tilde{\phi}$ neglected (the lines 5, 6, 7 giving the results in the first-, third- and fifth-order approximations, respectively). In our Tables 8.10(g–i) the second line gives results calculated by Luc-Koenig and Bachelier and quoted as private communication (1983) in Tables 2 and 4 in Korsch and Möhlenkamp (1983), and the lines 3–8 give our phase-integral results both with $\tilde{\phi}$ included (the lines 3, 4, 5 giving the results in the first-, third- and fifth-order approximations, respectively), and with $\tilde{\phi}$ neglected (the lines 6, 7, 8 giving the results in the first-, third- and fifth-order approximations, respectively).

For large values of n_2 (highly excited states of the η-equation) one can in general obtain more accurate values of E by the phase-integral method than by the numerical method used by Luc-Koenig and Bachelier (1980a), but for small values of n_2 the numerical method is in general more accurate.

Table 8.10(a). $F = 0.5 \times 10^{-5}$au, $|m| = 1, n = 14$.

n_1	n_2	$-E \times 10^3$	Γ	K	Z_1
0	12	4.00	1.1×10^{-3}		
		3.933 6	4.37×10^{-3}	−6.732	0.089 55
		3.933 327 7	$4.390\,5 \times 10^{-3}$	−6.738 2	0.089 626 37
		3.933 327 52	$4.390\,5 \times 10^{-3}$	−6.738 2	0.089 626 384
		3.935	4.31×10^{-3}	−6.70	0.089 57
		3.933 35	$4.390\,1 \times 10^{-3}$	−6.737 7	0.089 626 6
		3.933 330	$4.390\,5 \times 10^{-3}$	−6.738 2	0.089 626 41
1	11	3.80	0.84×10^{-3}		
		3.765 6	2.37×10^{-3}	−5.723	0.177 31
		3.765 448 2	$2.382\,2 \times 10^{-3}$	−5.728 6	0.177 380 466
		3.765 448 08	$2.382\,2 \times 10^{-3}$	−5.728 6	0.177 380 478
		3.767	2.34×10^{-3}	−5.69	0.177 35
		3.765 49	$2.382\,0 \times 10^{-3}$	−5.727 8	0.177 381 3
		3.765 454	$2.382\,2 \times 10^{-3}$	−5.728 4	0.177 380 6
2	10	3.61	0.70×10^{-3}		
		3.590 4	1.322×10^{-3}	−4.770	0.262 96
		3.590 311 2	$1.325\,8 \times 10^{-3}$	−4.775 4	0.263 028 027
		3.590 311 09	$1.325\,8 \times 10^{-3}$	−4.775 4	0.263 028 036
		3.592	1.30×10^{-3}	−4.73	0.263 022
		3.590 38	$1.325\,7 \times 10^{-3}$	−4.774	0.263 030
		3.590 325	$1.325\,8 \times 10^{-3}$	−4.775 2	0.263 028 5
3	9	3.41	4.2×10^{-4}		
		3.407 66	7.49×10^{-4}	−3.874	0.346 25
		3.407 571 4	$7.516\,2 \times 10^{-4}$	−3.878 9	0.346 319 002
		3.407 571 21	$7.516\,2 \times 10^{-4}$	−3.878 9	0.346 319 007
		3.410	7.38×10^{-4}	−3.83	0.346 35
		3.407 7	$7.516\,5 \times 10^{-4}$	−3.876	0.346 324
		3.407 60	$7.516\,9 \times 10^{-4}$	−3.878 2	0.346 320
4	8	3.219	3.1×10^{-4}		
		3.216 61	4.29×10^{-4}	−3.039	0.426 90
		3.216 580 8	$4.301\,4 \times 10^{-4}$	−3.043 8	0.426 969 943
		3.216 580 65	$4.301\,3 \times 10^{-4}$	−3.043 8	0.426 969 944
		3.219	4.22×10^{-4}	−2.99	0.427 04
		3.216 8	4.303×10^{-4}	−3.039	0.426 98
		3.216 64	$4.301\,6 \times 10^{-4}$	−3.042 5	0.426 973
5	7	3.017	1.9×10^{-4}		
		3.016 50	2.439×10^{-4}	−2.273	0.504 59
		3.016 523 4	$2.446\,5 \times 10^{-4}$	−2.277 2	0.504 654 997

(Continued)

Table 8.10(a). (*Continued*)

n_1	n_2	$-E \times 10^3$	Γ	K	Z_1
		3.016 523 19	$2.446\,5 \times 10^{-4}$	−2.277 2	0.504 654 991
		3.019	2.40×10^{-4}	−2.21	0.504 8
		3.016 9	$2.447\,1 \times 10^{-4}$	−2.269	0.504 68
		3.016 55	2.442×10^{-4}	−2.276 6	0.504 657
6	6	2.807	1.1×10^{-4}		
		2.806 7	1.340×10^{-4}	−1.576	0.578 95
		2.806 809 1	$1.344\,8 \times 10^{-4}$	−1.579 4	0.579 025 82
		2.806 808 72	$1.344\,8 \times 10^{-4}$	−1.579 4	0.579 025 799
		2.810	1.31×10^{-4}	−1.50	0.579 2
		2.807 2	1.337×10^{-4}	−1.571	0.579 06
		2.806 3	1.32×10^{-4}	−1.59	0.578 99
7	5	2.588	5.8×10^{-5}		
		2.587 5	6.78×10^{-5}	−0.928	0.649 69
		2.587 625 1	$6.814\,6 \times 10^{-5}$	−0.931 53	0.649 772 38
		2.587 624 56	$6.814\,5 \times 10^{-5}$	−0.931 52	0.649 772 338
		2.591	6.4×10^{-5}	−0.84	0.650 1
		2.587 1	6.5×10^{-5}	−0.94	0.649 73
		2.584	6.1×10^{-5}	−1.00	0.649 5
8	4	2.360 2	2.7×10^{-5}		
		2.360 1	2.94×10^{-5}	−0.281	0.716 62
		2.360 299 1	$2.960\,6 \times 10^{-5}$	−0.285 38	0.716 703 86
		2.360 298 3	$2.960\,5 \times 10^{-5}$	−0.285 37	0.716 703 784
		2.363	2.4×10^{-5}	−0.22	0.716 9
		2.355	2.2×10^{-5}	−0.4	0.716 1
		2.346	1.8×10^{-5}	−0.6	0.715
9	3	2.127 5	9.85×10^{-6}		
		2.127 2	9.7×10^{-6}	0.451	0.779 75
		2.127 484	$9.825\,9 \times 10^{-6}$	0.445 32	0.779 837 4
		2.127 482 7	$9.825\,4 \times 10^{-6}$	0.445 35	0.779 837 22
		2.123	10.5×10^{-6}	0.36	0.779 3
		2.130	6.8×10^{-6}	0.51	0.780 1
		2.136	3.6×10^{-6}	0.6	0.780 8
10	2	1.893 10	1.92×10^{-6}		
		1.892 8	1.93×10^{-6}	1.40	0.839 37
		1.893 098	$1.966\,4 \times 10^{-6}$	1.394 1	0.839 472 0
		1.893 095 6	$1.966\,1 \times 10^{-6}$	1.394 2	0.839 471 77
		1.889	2.3×10^{-6}	1.32	0.838 9
		1.892 6	2.00×10^{-6}	1.38	0.839 42
		1.893 5	1.8×10^{-6}	1.40	0.839 53

(*Continued*)

Table 8.10(a). (*Continued*)

n_1	n_2	$-E \times 10^3$	Γ	K	Z_1
11	1	1.660 368	1.674×10^{-7}		
		1.660 1	1.62×10^{-7}	2.74	0.895 897
		1.660 372	1.674×10^{-7}	2.722 3	0.896 000 7
		1.660 368 0	1.672×10^{-7}	2.722 7	0.896 000 12
		1.658	1.8×10^{-7}	2.69	0.895 6
		1.660 2	1.70×10^{-7}	2.717	0.895 97
		1.660 31	1.678×10^{-7}	2.721 5	0.895 993
12	0	1.429 565 4	4.086×10^{-9}		
		1.429 2	3.7×10^{-9}	4.69	0.949 4
		1.429 58	4.13×10^{-9}	4.635	0.949 531
		1.429 561	4.046×10^{-9}	4.645 3	0.949 528 4
		1.428	4.0×10^{-9}	4.65	0.949 2
		1.429 51	4.15×10^{-9}	4.633	0.949 521
		1.429 557	4.048×10^{-9}	4.645 2	0.949 527 9

Korsch and Möhlenkamp (1983) have used semiclassical methods with real energy and with complex energy for resonant states corresponding to states in our Tables 8.10(g–i). Their energy values are in satisfactory agreement with the first-order approximation, with $\tilde{\phi}$ retained, of our phase-integral results. The assertion in their Table 4 that states with $n_1 \leq 8$ for $n = 11$ lie above the barrier is not in complete agreement with our results in Table 8.10(g), since K is negative for the state with $n_1 = 8$.

In Tables 8.10(c), 8.10(f) and 8.10(l) there are some states with very large K-values for which the numerical method gives only an upper limit for Γ, while the phase-integral method gives a good value of Γ.

Table 8.10(b). $F = 0.5 \times 10^{-5}$au, $|m| = 1, n = 13$.

n_1	n_2	$-E \times 10^3$	Γ	K	Z_1
0	11	4.224	5.0×10^{-5}		
		4.224 07	5.68×10^{-5}	−0.717	0.092 71
		4.224 056 0	$5.693 2 \times 10^{-5}$	−0.719 18	0.092 783 57
		4.224 055 89	$5.693 2 \times 10^{-5}$	−0.719 18	0.092 783 58
		4.228	5.15×10^{-5}	−0.63	0.092 76
		4.223	5.20×10^{-5}	−0.75	0.092 77
		4.218	4.6×10^{-5}	−0.84	0.092 72
1	10	4.025 4	3.0×10^{-5}		
		4.025 568	3.36×10^{-5}	−0.320	0.182 97
		4.025 569 4	$3.369 8 \times 10^{-5}$	−0.321 84	0.183 042 117
		4.025 569 21	$3.369 8 \times 10^{-5}$	−0.321 84	0.183 042 126
		4.029	2.7×10^{-5}	−0.25	0.183 04
		4.020	2.6×10^{-5}	−0.44	0.182 9
		4.012	2.1×10^{-5}	−0.6	0.182 7
2	9	3.824 0	1.7×10^{-5}		
		3.823 99	1.813×10^{-5}	0.106	0.270 61
		3.824 009 4	$1.819 5 \times 10^{-5}$	0.103 37	0.270 677 211
		3.824 009 26	$1.819 5 \times 10^{-5}$	0.103 37	0.270 677 216
		3.822	1.4×10^{-5}	0.06	0.270 5
		3.833	0.7×10^{-5}	0.3	0.271 0
		3.84	0.3×10^{-5}	0.5	0.271 3
3	8	3.620 3	8.34×10^{-6}		
		3.620 25	8.56×10^{-6}	0.587	0.355 56
		3.620 276 6	$8.599 3 \times 10^{-6}$	0.584 01	0.355 628 623
		3.620 276 35	$8.599 3 \times 10^{-6}$	0.584 01	0.355 628 624
		3.616	9.8×10^{-6}	0.49	0.355 4
		3.622	6.7×10^{-6}	0.62	0.355 71
		3.627	4.1×10^{-6}	0.73	0.355 9
4	7	3.415 42	3.31×10^{-6}		
		3.415 40	3.31×10^{-6}	1.157	0.437 82
		3.415 426 0	$3.327 4 \times 10^{-6}$	1.153 1	0.437 883 566
		3.415 425 72	$3.327 4 \times 10^{-6}$	1.153 1	0.437 883 562
		3.411	4.1×10^{-6}	1.06	0.437 6
		3.415 1	3.29×10^{-6}	1.147	0.437 87
		3.417	2.9×10^{-6}	1.18	0.437 95
5	6	3.210 36	9.616×10^{-7}		
		3.210 33	9.68×10^{-7}	1.848	0.517 40
		3.210 363 2	$9.755 3 \times 10^{-7}$	1.843 4	0.517 466 23
		3.210 362 87	$9.755 3 \times 10^{-7}$	1.843 4	0.517 466 222
		3.207	11.5×10^{-7}	1.78	0.517 2
		3.209 9	10.0×10^{-7}	1.83	0.517 44
		3.210 40	9.6×10^{-7}	1.844	0.517 468

(*Continued*)

Table 8.10(b). (*Continued*)

n_1	n_2	$-E \times 10^3$	Γ	K	Z_1
6	5	3.005 480	2.049×10^{-7}		
		3.005 44	2.03×10^{-7}	2.687	0.594 33
		3.005 479 3	$2.052\,2 \times 10^{-7}$	2.681 5	0.594 402 36
		3.005 478 93	$2.052\,1 \times 10^{-7}$	2.681 5	0.594 402 342
		3.003	2.3×10^{-7}	2.63	0.594 1
		3.005 2	2.08×10^{-7}	2.676	0.594 38
		3.005 42	2.059×10^{-7}	2.680	0.594 398
7	4	2.800 685 5	$2.992\,3 \times 10^{-8}$		
		2.800 64	2.95×10^{-8}	3.696	0.668 63
		2.800 685 9	$2.993\,0 \times 10^{-8}$	3.689 7	0.668 693 99
		2.800 685 47	$2.992\,9 \times 10^{-8}$	3.689 7	0.668 693 959
		2.799	3.2×10^{-8}	3.66	0.668 5
		2.800 58	3.01×10^{-8}	3.688	0.668 685
		2.800 66	2.998×10^{-8}	3.689 2	0.668 692
8	3	2.595 758 92	$2.883\,0 \times 10^{-9}$		
		2.595 69	2.83×10^{-9}	4.907	0.740 26
		2.595 759 5	$2.883\,0 \times 10^{-9}$	4.897 7	0.740 326 73
		2.595 758 92	$2.882\,8 \times 10^{-9}$	4.897 8	0.740 326 674
		2.594	3.0×10^{-9}	4.87	0.740 1
		2.595 72	2.890×10^{-9}	4.896 7	0.740 322
		2.595 752	2.884×10^{-9}	4.897 6	0.740 326 0
9	2	2.390 542 956	$1.672\,1 \times 10^{-10}$		
		2.390 46	1.63×10^{-10}	6.37	0.809 21
		2.390 544	$1.672\,5 \times 10^{-10}$	6.354 7	0.809 284 35
		2.390 542 94	$1.672\,1 \times 10^{-10}$	6.354 8	0.809 284 255
		2.389 3	1.73×10^{-10}	6.34	0.809 1
		2.390 52	1.674×10^{-10}	6.354 2	0.809 282
		2.390 541	$1.672\,3 \times 10^{-10}$	6.354 8	0.809 284 1
10	1	2.184 962 854	$4.899\,0 \times 10^{-12}$		
		2.184 86	4.7×10^{-12}	8.17	0.875 48
		2.184 965	4.903×10^{-12}	8.149 8	0.875 554 3
		2.184 962 79	$4.898\,6 \times 10^{-12}$	8.150 2	0.875 554 09
		2.183 9	4.93×10^{-12}	8.148	0.875 4
		2.184 95	4.906×10^{-12}	8.149 5	0.875 553
		2.184 962 3	$4.898\,7 \times 10^{-12}$	8.150 2	0.875 554 04
11	0	1.978 993 997	$4.841\,8 \times 10^{-14}$		
		1.978 85	4.4×10^{-14}	10.53	0.939 05
		1.979 000	4.906×10^{-14}	10.480	0.939 128 4
		1.978 992 1	$4.804\,9 \times 10^{-14}$	10.490	0.939 127 35
		1.978 07	4.6×10^{-14}	10.51	0.938 9
		1.978 988	4.909×10^{-14}	10.480	0.939 126 8
		1.978 993 8	$4.804\,6 \times 10^{-14}$	10.490	0.939 127 58

Table 8.10(c). $F = 0.5 \times 10^{-5}$au, $|m| = 1, n = 12$.

n_1	n_2	$-E \times 10^3$	Γ	K	Z_1
0	10	4.466 560 3	8.193×10^{-9}		
		4.466 7	8.12×10^{-9}	4.417	0.095 27
		4.466 560 5	$8.192 9 \times 10^{-9}$	4.413 0	0.095 339 219
		4.466 560 346	$8.192 9 \times 10^{-9}$	4.413 0	0.095 339 228
		4.465	8.8×10^{-9}	4.38	0.095 26
		4.466 50	8.22×10^{-9}	4.412	0.095 338 6
		4.466 55	8.199×10^{-9}	4.412 7	0.095 339 1
1	9	4.285 917 3	1.775×10^{-9}		
		4.286 0	1.76×10^{-9}	5.208	0.188 48
		4.285 917 4	$1.774 7 \times 10^{-9}$	5.203 2	0.188 547 471
		4.285 917 279	$1.774 7 \times 10^{-9}$	5.203 2	0.188 547 478
		4.284	1.9×10^{-9}	5.18	0.188 45
		4.285 88	1.778×10^{-9}	5.202	0.188 546 7
		4.285 912	$1.775 3 \times 10^{-9}$	5.203 1	0.188 547 37
2	8	4.104 666 30	$3.221 8 \times 10^{-10}$		
		4.104 8	3.19×10^{-10}	6.085	0.279 55
		4.104 666 4	$3.221 8 \times 10^{-10}$	6.080 2	0.279 608 726
		4.104 666 303	$3.221 8 \times 10^{-10}$	6.080 2	0.279 608 731
		4.103	3.4×10^{-10}	6.06	0.279 50
		4.104 64	3.226×10^{-10}	6.079 7	0.279 608 0
		4.104 664	$3.222 3 \times 10^{-10}$	6.080 2	0.279 608 66
3	7	3.922 792 971	$4.799 7 \times 10^{-11}$		
		3.922 9	4.75×10^{-11}	7.06	0.368 45
		3.922 793 1	$4.799 8 \times 10^{-11}$	7.054 8	0.368 506 601
		3.922 792 971	$4.799 8 \times 10^{-11}$	7.054 8	0.368 506 603
		3.921 7	5.0×10^{-11}	7.03	0.368 40
		3.922 78	4.804×10^{-11}	7.054 5	0.368 505 9
		3.922 791 9	$4.800 1 \times 10^{-11}$	7.054 8	0.368 506 56
4	6	3.740 291 621	$5.708 9 \times 10^{-12}$		
		3.740 4	5.64×10^{-12}	8.15	0.455 17
		3.740 291 8	$5.709 0 \times 10^{-12}$	8.140 9	0.455 224 808
		3.740 291 621	$5.709 0 \times 10^{-12}$	8.140 9	0.455 224 806
		3.739	5.9×10^{-12}	8.12	0.455 11
		3.740 28	5.712×10^{-12}	8.140 7	0.455 224 2
		3.740 291 1	$5.709 1 \times 10^{-12}$	8.140 9	0.455 224 78
5	5	3.557 163 134	$5.229 8 \times 10^{-13}$		
		3.557 23	5.16×10^{-13}	9.364	0.539 69
		3.557 163 4	$5.229 9 \times 10^{-13}$	9.356 7	0.539 747 481
		3.557 163 134	$5.229 8 \times 10^{-13}$	9.356 7	0.539 747 474

(*Continued*)

Table 8.10(c). *(Continued)*

n_1	n_2	$-E \times 10^3$	Γ	K	Z_1
		3.556	5.4×10^{-13}	9.34	0.539 63
		3.557 156	5.232×10^{-13}	9.356 5	0.539 747 0
		3.557 162 9	$5.229 8 \times 10^{-13}$	9.356 7	0.539 747 46
6	4	3.373 413 570	$3.514 0 \times 10^{-14}$		
		3.373 47	3.46×10^{-14}	10.735	0.622 00
		3.373 413 9	$3.514 4 \times 10^{-14}$	10.727	0.622 059 43
		3.373 413 569	$3.514 3 \times 10^{-14}$	10.727	0.622 059 416
		3.372 6	3.6×10^{-14}	10.715	0.621 9
		3.373 408	$3.515 4 \times 10^{-14}$	10.726	0.622 059 0
		3.373 413 4	$3.514 3 \times 10^{-14}$	10.727	0.622 059 406
7	3	3.189 053 369	$1.628 0 \times 10^{-15}$		
		3.189 09	1.58×10^{-15}	12.296	0.702 09
		3.189 053 7	$1.616 7 \times 10^{-15}$	12.285	0.702 146 36
		3.189 053 366	$1.616 6 \times 10^{-15}$	12.285	0.702 146 337
		3.188 3	1.64×10^{-15}	12.278	0.702 02
		3.189 050	$1.617 1 \times 10^{-15}$	12.285	0.702 146 0
		3.189 053 29	$1.616 6 \times 10^{-15}$	12.285	0.702 146 331
8	2	3.004 096 870	$4.648 0 \times 10^{-17}$		
		3.004 12	4.5×10^{-17}	14.10	0.779 94
		3.004 097 4	$4.581 0 \times 10^{-17}$	14.086	0.779 995 08
		3.004 096 86	$4.580 1 \times 10^{-17}$	14.086	0.779 995 033
		3.003	4.61×10^{-17}	14.083	0.779 87
		3.004 094	$4.581 8 \times 10^{-17}$	14.086	0.779 994 8
		3.004 096 83	$4.580 1 \times 10^{-17}$	14.086	0.779 995 030
9	1	2.818 562 001	$< 10^{-17}$		
		2.818 568	6.4×10^{-19}	16.24	0.855 54
		2.818 563 0	6.662×10^{-19}	16.219	0.855 593 68
		2.818 561 97	6.656×10^{-19}	16.219	0.855 593 577
		2.817 9	6.58×10^{-19}	16.225	0.855 47
		2.818 560	6.663×10^{-19}	16.219	0.855 593 4
		2.818 561 98	$6.656 0 \times 10^{-19}$	16.219	0.855 593 578
10	0	2.632 470 066	$< 10^{-17}$		
		2.632 45	3.0×10^{-21}	18.94	0.928 87
		2.632 474	3.346×10^{-21}	18.883	0.928 932 0
		2.632 468 9	$3.277 6 \times 10^{-21}$	18.893	0.928 931 38
		2.631 9	3.1×10^{-21}	18.92	0.928 80
		2.632 468 6	3.347×10^{-21}	18.883	0.928 931 3
		2.632 470 05	$3.277 5 \times 10^{-21}$	18.893	0.928 931 51

Table 8.10(d). $F = 1.0 \times 10^{-5}$au, $|m| = 1, n = 12$.

n_1	n_2	$-E \times 10^3$	Γ	K	Z_1
0	10	5.53	0.20×10^{-2}		
		5.354 5	1.276×10^{-2}	−7.483	0.104 74
		5.353 898 2	$1.282\,7 \times 10^{-2}$	−7.491 5	0.104 840 57
		5.353 897 81	$1.282\,7 \times 10^{-2}$	−7.491 5	0.104 840 605
		5.357	1.26×10^{-2}	−7.45	0.104 76
		5.353 93	$1.282\,6 \times 10^{-2}$	−7.491 1	0.104 840 9
		5.353 900	$1.282\,7 \times 10^{-2}$	−7.491 5	0.104 840 63
1	9	5.19	1.6×10^{-3}		
		5.087 5	6.06×10^{-3}	−6.274	0.207 23
		5.087 052 6	$6.093\,0 \times 10^{-3}$	−6.281 9	0.207 327 01
		5.087 052 29	$6.093\,0 \times 10^{-3}$	−6.281 9	0.207 327 039
		5.090	5.97×10^{-3}	−6.24	0.207 29
		5.087 11	$6.092\,3 \times 10^{-3}$	−6.281 2	0.207 328 0
		5.087 058	$6.093\,0 \times 10^{-3}$	−6.281 8	0.207 327 16
2	8	4.84	1.2×10^{-3}		
		4.803 8	3.02×10^{-3}	−5.154	0.306 91
		4.803 460 0	$3.035\,0 \times 10^{-3}$	−5.160 8	0.306 997 81
		4.803 459 64	$3.035\,0 \times 10^{-3}$	−5.160 8	0.306 997 832
		4.807	2.97×10^{-3}	−5.11	0.306 996
		4.803 55	$3.034\,6 \times 10^{-3}$	−5.159 6	0.307 000
		4.803 48	$3.035\,0 \times 10^{-3}$	−5.160 6	0.306 998 3
3	7	4.52	1.0×10^{-3}		
		4.503 2	1.556×10^{-3}	−4.112	0.403 30
		4.503 042 7	$1.563\,1 \times 10^{-3}$	−4.118 9	0.403 389 428
		4.503 042 35	$1.563\,1 \times 10^{-3}$	−4.118 9	0.403 389 437
		4.507	1.53×10^{-3}	−4.07	0.403 43
		4.503 2	$1.563\,0 \times 10^{-3}$	−4.117	0.403 396
		4.503 09	$1.563\,2 \times 10^{-3}$	−4.118 3	0.403 391
4	6	4.19	6.5×10^{-4}		
		4.184 73	8.17×10^{-4}	−3.152	0.495 89
		4.184 655 2	$8.202\,6 \times 10^{-4}$	−3.157 5	0.495 983 454
		4.184 654 70	$8.202\,6 \times 10^{-4}$	−3.157 5	0.495 983 451 0
		4.189	8.0×10^{-4}	−3.10	0.496 09
		4.185 0	8.205×10^{-4}	−3.153	0.496 001
		4.184 76	$8.203\,3 \times 10^{-4}$	−3.156	0.495 988
5	5	3.85	3.2×10^{-4}		
		3.846 35	4.27×10^{-4}	−2.278	0.584 07
		3.846 424 5	$4.288\,0 \times 10^{-4}$	−2.283 4	0.584 172 58

(*Continued*)

Table 8.10(d). (*Continued*)

n_1	n_2	$-E \times 10^3$	Γ	K	Z_1
		3.846 423 73	4.2880×10^{-4}	−2.283 4	0.584 172 558
		3.851	4.19×10^{-4}	−2.21	0.584 4
		3.847 1	4.2889×10^{-4}	−2.275	0.584 21
		3.846 47	4.2805×10^{-4}	−2.282 9	0.584 175
6	4	3.487	1.7×10^{-4}		
		3.486 6	2.13×10^{-4}	−1.490	0.667 18
		3.486 836 4	2.1401×10^{-4}	−1.495 0	0.667 294 16
		3.486 834 98	2.1401×10^{-4}	−1.494 9	0.667 294 084
		3.493	2.07×10^{-4}	−1.41	0.667 6
		3.487 4	2.12×10^{-4}	−1.487	0.667 33
		3.485 7	2.09×10^{-4}	−1.51	0.667 22
7	3	3.106	8.2×10^{-5}		
		3.106 0	9.36×10^{-5}	−0.752	0.744 65
		3.106 383	9.4336×10^{-5}	−0.757 67	0.744 769 2
		3.106 380 2	9.4332×10^{-5}	−0.757 64	0.744 768 99
		3.113	8.6×10^{-5}	−0.66	0.745 2
		3.104	8.7×10^{-5}	−0.79	0.744 6
		3.098	7.8×10^{-5}	−0.87	0.744 1
8	2	2.708 7	3.0×10^{-5}		
		2.708 2	3.17×10^{-5}	0.027	0.816 2
		2.708 779	3.2100×10^{-5}	0.018 89	0.816 304 6
		2.708 772 9	3.2095×10^{-5}	0.018 98	0.816 304 10
		2.706 7	1.8×10^{-5}	0.006	0.816 0
		2.73	1.0×10^{-5}	0.2	0.818
		2.74	0.4×10^{-5}	0.5	0.819
9	1	2.302 0	6.36×10^{-6}		
		2.301 3	6.2×10^{-6}	1.03	0.882 0
		2.302 070	6.351×10^{-6}	1.018 8	0.882 166 3
		2.302 056 7	6.3461×10^{-6}	1.019 2	0.882 165 08
		2.294	7.6×10^{-6}	0.94	0.881 3
		2.301 8	6.2×10^{-6}	1.015	0.882 14
		2.305	5.1×10^{-6}	1.06	0.882 4
10	0	1.896 981	3.632×10^{-7}		
		1.896 1	3.3×10^{-7}	2.62	0.943 1
		1.897 03	3.68×10^{-7}	2.576	0.943 238
		1.896 962	3.596×10^{-7}	2.586 6	0.943 231 4
		1.893	3.8×10^{-7}	2.569 1	0.942 7
		1.896 6	3.75×10^{-7}	2.568 5	0.943 19
		1.896 88	3.609×10^{-7}	2.585 2	0.943 222

Table 8.10(e). $F = 1.0 \times 10^{-5}$au, $|m| = 1, n = 11$.

n_1	n_2	$-E \times 10^3$	Γ	K	Z_1
0	9	5.899	0.94×10^{-4}		
		5.899 28	1.109×10^{-4}	−0.807	0.109 76
		5.899 246 4	$1.113 3 \times 10^{-4}$	−0.810 23	0.109 861 57
		5.899 246 02	$1.113 3 \times 10^{-4}$	−0.810 23	0.109 861 594
		5.907	1.02×10^{-4}	−0.72	0.109 83
		5.897	1.04×10^{-4}	−0.83	0.109 85
		5.891	0.94×10^{-4}	−0.91	0.109 79
1	8	5.565	5.8×10^{-5}		
		5.564 987	6.44×10^{-5}	−0.390	0.216 05
		5.564 994 9	$6.468 9 \times 10^{-5}$	−0.393 33	0.216 142 608
		5.564 994 50	$6.468 9 \times 10^{-5}$	−0.393 33	0.216 142 623
		5.571	5.3×10^{-5}	−0.31	0.216 16
		5.557	5.2×10^{-5}	−0.49	0.216 0
		5.54	4.2×10^{-5}	−0.7	0.215 8
2	7	5.224 1	3.3×10^{-5}		
		5.224 12	3.38×10^{-5}	0.058	0.318 57
		5.224 165 2	$3.393 6 \times 10^{-5}$	0.054 872	0.318 665 598
		5.224 164 65	$3.393 6 \times 10^{-5}$	0.054 874	0.318 665 605
		5.222	2.2×10^{-5}	0.03	0.318 50
		5.24	1.1×10^{-5}	0.3	0.319 1
		5.26	0.5×10^{-5}	0.5	0.319 6
3	6	4.878 5	1.4×10^{-5}		
		4.878 42	1.516×10^{-5}	0.573	0.417 22
		4.878 497 1	$1.525 4 \times 10^{-5}$	0.569 37	0.417 314 925
		4.878 496 41	$1.525 4 \times 10^{-5}$	0.569 37	0.417 314 920
		4.871	1.7×10^{-5}	0.47	0.416 9
		4.882	1.2×10^{-5}	0.61	0.417 43
		4.891	0.7×10^{-5}	0.72	0.417 8
4	5	4.530 2	5.32×10^{-6}		
		4.530 14	5.32×10^{-6}	1.200	0.511 96
		4.530 232 2	$5.361 5 \times 10^{-6}$	1.195 6	0.512 061 57
		4.530 231 36	$5.361 5 \times 10^{-6}$	1.195 6	0.512 061 545
		4.523	6.6×10^{-6}	1.10	0.511 6
		4.529 6	5.33×10^{-6}	1.188	0.512 03
		4.532	4.7×10^{-6}	1.22	0.512 15
5	4	4.181 42	1.31×10^{-6}		
		4.181 31	1.297×10^{-6}	1.989	0.602 85
		4.181 425 3	$1.311 5 \times 10^{-6}$	1.983 4	0.602 950 36

(*Continued*)

Table 8.10(e). (*Continued*)

n_1	n_2	$-E \times 10^3$	Γ	K	Z_1
		4.181 424 25	$1.311\,5 \times 10^{-6}$	1.983 4	0.602 950 320
		4.176	1.5×10^{-6}	1.92	0.602 6
		4.180 7	1.34×10^{-6}	1.974	0.602 91
		4.181 39	1.305×10^{-6}	1.983 0	0.602 949
6	3	3.832 943	2.036×10^{-7}		
		3.832 81	2.00×10^{-7}	2.986	0.689 93
		3.832 944 3	$2.033\,6 \times 10^{-7}$	2.977 8	0.690 028 14
		3.832 942 87	$2.033\,4 \times 10^{-7}$	2.977 9	0.690 028 058
		3.829	2.2×10^{-7}	2.94	0.689 7
		3.832 6	2.06×10^{-7}	2.974	0.690 01
		3.832 86	2.040×10^{-7}	2.976 8	0.690 023
7	2	3.484 550 8	$1.835\,6 \times 10^{-8}$		
		3.484 38	1.79×10^{-8}	4.24	0.773 19
		3.484 553	$1.835\,7 \times 10^{-8}$	4.228 7	0.773 294 5
		3.484 550 8	$1.835\,4 \times 10^{-8}$	4.228 8	0.773 294 30
		3.482	1.95×10^{-8}	4.20	0.773 0
		3.484 43	1.844×10^{-8}	4.227	0.773 285
		3.484 53	1.837×10^{-8}	4.228 5	0.773 292 5
8	1	3.135 804 72	$8.233\,9 \times 10^{-10}$		
		3.135 6	7.9×10^{-10}	5.84	0.852 62
		3.135 809	8.241×10^{-10}	5.821 0	0.852 724 4
		3.135 804 5	$8.233\,3 \times 10^{-10}$	5.821 5	0.852 724 07
		3.133	8.4×10^{-10}	5.81	0.852 4
		3.135 76	8.26×10^{-10}	5.820 3	0.852 720
		3.135 800	8.235×10^{-10}	5.821 4	0.852 723 7
9	0	2.786 465 882	$1.239\,9 \times 10^{-11}$		
		2.786 2	1.13×10^{-11}	8.00	0.928 18
		2.786 48	1.255×10^{-11}	7.949 2	0.928 298
		2.786 461	$1.228\,8 \times 10^{-11}$	7.959 6	0.928 295 9
		2.784 5	1.19×10^{-11}	7.98	0.928 0
		2.786 44	1.256×10^{-11}	7.948 6	0.928 294
		2.786 464 8	$1.228\,8 \times 10^{-11}$	7.959 7	0.928 296 2

Table 8.10(f). $F = 1.0 \times 10^{-5}$au, $|m| = 1, n = 10$.

n_1	n_2	$-E \times 10^3$	Γ	K	Z_1
0	8	6.324 857	$1.116\,2 \times 10^{-8}$		
		6.325 1	1.10×10^{-8}	4.559	0.113 54
		6.324 857 1	$1.116\,2 \times 10^{-8}$	4.553 4	0.113 630 87
		6.324 856 720	$1.116\,2 \times 10^{-8}$	4.553 4	0.113 630 891
		6.322	1.20×10^{-8}	4.52	0.113 51
		6.324 75	1.120×10^{-8}	4.552	0.113 630 0
		6.324 84	$1.117\,0 \times 10^{-8}$	4.553 2	0.113 630 7
1	7	6.023 885 0	$2.122\,6 \times 10^{-9}$		
		6.024 1	2.10×10^{-9}	5.417	0.224 21
		6.023 885 4	$2.122\,5 \times 10^{-9}$	5.410 7	0.224 301 271
		6.023 885 043	$2.122\,5 \times 10^{-9}$	5.410 7	0.224 301 283
		6.021	2.3×10^{-9}	5.38	0.224 17
		6.023 82	2.127×10^{-9}	5.409 9	0.224 300 2
		6.023 877	$2.123\,2 \times 10^{-9}$	5.410 6	0.224 301 1
2	6	5.721 681 79	$3.234\,3 \times 10^{-10}$		
		5.721 9	3.19×10^{-10}	6.384	0.331 90
		5.721 682 2	$3.234\,4 \times 10^{-10}$	6.377 0	0.331 984 207
		5.721 681 789	$3.234\,3 \times 10^{-10}$	6.377 0	0.331 984 212
		5.720	3.4×10^{-10}	6.35	0.331 8
		5.721 64	3.238×10^{-10}	6.376 5	0.331 983 2
		5.721 678	$3.234\,7 \times 10^{-10}$	6.376 9	0.331 984 12
3	5	5.418 233 323	$3.812\,0 \times 10^{-11}$		
		5.418 4	3.76×10^{-11}	7.478	0.436 57
		5.418 233 8	$3.812\,7 \times 10^{-11}$	7.470 3	0.436 652 674
		5.418 233 321	$3.812\,7 \times 10^{-11}$	7.470 3	0.436 652 671
		5.416	4.0×10^{-11}	7.45	0.436 50
		5.418 21	3.816×10^{-11}	7.470 0	0.436 651 8
		5.418 232	$3.812\,9 \times 10^{-11}$	7.470 3	0.436 652 62
4	4	5.113 542 861	$3.312\,9 \times 10^{-12}$		
		5.113 7	3.26×10^{-12}	8.724	0.538 20
		5.113 543 5	$3.313\,0 \times 10^{-12}$	8.715 1	0.538 280 375
		5.113 542 859	$3.312\,9 \times 10^{-12}$	8.715 1	0.538 280 361
		5.112	3.42×10^{-12}	8.70	0.538 1
		5.113 526	3.315×10^{-12}	8.714 9	0.538 279 6
		5.113 542 2	$3.312\,9 \times 10^{-12}$	8.715 1	0.538 280 33
5	3	4.807 626 564	$1.980\,4 \times 10^{-13}$		
		4.807 74	1.94×10^{-13}	10.16	0.636 76
		4.807 627 4	$1.980\,6 \times 10^{-13}$	10.146	0.636 842 28

(*Continued*)

Table 8.10(f). (*Continued*)

n_1	n_2	$-E \times 10^3$	Γ	K	Z_1
		4.807 626 558	$1.980\,4 \times 10^{-13}$	10.146	0.636 842 250
		4.806	2.03×10^{-13}	10.135	0.636 68
		4.807 615	$1.981\,3 \times 10^{-13}$	10.146	0.636 841 6
		4.807 626 2	$1.980\,4 \times 10^{-13}$	10.146	0.636 842 23
6	2	4.500 511 277	$7.330\,9 \times 10^{-15}$		
		4.500 60	7.1×10^{-15}	11.83	0.732 24
		4.500 512 5	7.330×10^{-15}	11.816	0.732 315 09
		4.500 511 261	$7.328\,1 \times 10^{-15}$	11.816	0.732 315 029
		4.499	7.42×10^{-15}	11.810	0.732 1
		4.500 503	7.332×10^{-15}	11.815	0.732 314 5
		4.500 511 1	$7.328\,2 \times 10^{-15}$	11.816	0.732 315 020
7	1	4.192 233 275	1.200×10^{-16}		
		4.192 28	1.34×10^{-16}	13.84	0.824 60
		4.192 235	$1.399\,1 \times 10^{-16}$	13.815	0.824 677 7
		4.192 233 20	$1.397\,8 \times 10^{-16}$	13.816	0.824 677 524
		4.191	1.389×10^{-16}	13.819	0.824 50
		4.192 228	$1.399\,4 \times 10^{-16}$	13.815	0.824 677 1
		4.192 233 19	$1.397\,8 \times 10^{-16}$	13.816	0.824 677 523
8	0	3.882 837 521	$< 10^{-17}$		
		3.882 831	8.29×10^{-19}	16.40	0.913 83
		3.882 846	9.267×10^{-19}	16.344	0.913 912
		3.882 835	$9.078\,3 \times 10^{-19}$	16.354	0.913 910 9
		3.881 7	8.6×10^{-19}	16.38	0.913 74
		3.882 834	9.270×10^{-19}	16.344	0.913 910 8
		3.882 837 48	$9.077\,9 \times 10^{-19}$	16.354	0.913 911 10

Table 8.10(g). $F = 1.5 \times 10^{-5}$au, $|m| = 1, n = 11$.

n_1	n_2	$-E \times 10^3$	Γ	K	Z_1
0	9	6.71	0.32×10^{-2}		
		6.705 9	0.32×10^{-2}		
		6.344	2.93×10^{-2}	−8.235	0.114 2
		6.342 663 8	$2.946\,8 \times 10^{-2}$	−8.246 8	0.114 346 29
		6.342 663 03	$2.946\,8 \times 10^{-2}$	−8.246 8	0.114 346 339
		6.348	2.88×10^{-2}	−8.20	0.114 25
		6.342 70	$2.946\,5 \times 10^{-2}$	−8.246 5	0.114 346 6
		6.342 665	$2.946\,8 \times 10^{-2}$	−8.246 8	0.114 346 36
1	8	6.19	0.20×10^{-2}		
		6.194 7	0.20×10^{-2}		
		6.003 4	1.220×10^{-2}	−6.832	0.226 07
		6.002 595 0	$1.227\,5 \times 10^{-2}$	−6.841 8	0.226 188 44
		6.002 594 35	$1.227\,5 \times 10^{-2}$	−6.841 8	0.226 188 481
		6.007	1.20×10^{-2}	−6.80	0.226 14
		6.002 65	$1.227\,3 \times 10^{-2}$	−6.841 3	0.226 189 5
		6.002 600	$1.227\,5 \times 10^{-2}$	−6.841 8	0.226 188 58
2	7	5.72	2.0×10^{-3}		
		5.719 4	2.0×10^{-3}		
		5.633 8	5.49×10^{-3}	−5.557	0.334 67
		5.633 211 7	$5.522\,6 \times 10^{-3}$	−5.566 1	0.334 784 56
		5.633 211 18	$5.522\,5 \times 10^{-3}$	−5.566 1	0.334 784 590
		5.638	5.40×10^{-3}	−5.52	0.334 786
		5.633 32	$5.521\,8 \times 10^{-3}$	−5.565 1	0.334 787
		5.633 23	$5.522\,6 \times 10^{-3}$	−5.565 9	0.334 785 0
3	6	5.26	1.4×10^{-3}		
		5.263 1	1.4×10^{-3}		
		5.236 3	2.60×10^{-3}	−4.387	0.439 34
		5.235 975 2	$2.615\,8 \times 10^{-3}$	−4.394 6	0.439 448 226
		5.235 974 66	$2.615\,8 \times 10^{-3}$	−4.394 6	0.439 448 241
		5.241	2.56×10^{-3}	−4.34	0.439 51
		5.236 2	$2.615\,6 \times 10^{-3}$	−4.393	0.439 455
		5.236 02	$2.615\,9 \times 10^{-3}$	−4.394 2	0.439 450
4	5	4.81	0.94×10^{-3}		
		4.813 7	0.94×10^{-3}		
		4.809 91	1.270×10^{-3}	−3.314	0.539 3
		4.809 776 6	$1.276\,8 \times 10^{-3}$	−3.321 0	0.539 432 301
		4.809 775 74	$1.276\,8 \times 10^{-3}$	−3.321 0	0.539 432 296
		4.816	1.25×10^{-3}	−3.26	0.539 6
		4.810 2	$1.277\,1 \times 10^{-3}$	−3.317	0.539 452
		4.809 90	$1.276\,9 \times 10^{-3}$	−3.319 8	0.539 438

(*Continued*)

Table 8.10(g). *(Continued)*

n_1	n_2	$-E \times 10^3$	Γ	K	Z_1
5	4	4.34	4.5×10^{-4}		
		4.345 9	4.5×10^{-4}		
		4.351 49	6.20×10^{-4}	−2.344	0.633 7
		4.351 625 3	$6.239\,5 \times 10^{-4}$	−2.350 6	0.633 848 59
		4.351 623 75	$6.239\,4 \times 10^{-4}$	−2.350 6	0.633 848 538
		4.358	6.1×10^{-4}	−2.28	0.634 1
		4.352 5	6.241×10^{-4}	−2.343	0.633 89
		4.351 71	6.231×10^{-4}	−2.349 8	0.633 853
6	3	3.86	2.4×10^{-4}		
		3.857 9	2.4×10^{-4}		
		3.858 2	2.87×10^{-4}	−1.474	0.721 56
		3.858 616	$2.889\,2 \times 10^{-4}$	−1.481 0	0.721 698 0
		3.858 613 1	$2.889\,1 \times 10^{-4}$	−1.480 9	0.721 697 85
		3.867	2.79×10^{-4}	−1.39	0.722 1
		3.859 4	2.86×10^{-4}	−1.474	0.721 74
		3.857 0	2.82×10^{-4}	−1.50	0.721 60
7	2	3.331	0.82×10^{-4}		
		3.330 8	0.82×10^{-4}		
		3.330 5	1.12×10^{-4}	−0.656	0.801 9
		3.331 311	$1.136\,6 \times 10^{-4}$	−0.664 15	0.802 101 5
		3.331 303 0	$1.136\,5 \times 10^{-4}$	−0.664 05	0.802 100 98
		3.339	1.01×10^{-4}	−0.57	0.802 6
		3.327	1.02×10^{-4}	−0.71	0.801 8
		3.317	0.9×10^{-4}	−0.81	0.801 1
8	1	2.777	2.9×10^{-5}		
		2.776 5	2.9×10^{-5}		
		2.775 4	3.00×10^{-5}	0.254	0.874 5
		2.776 64	3.069×10^{-5}	0.240 5	0.874 682
		2.776 621	$3.066\,9 \times 10^{-5}$	0.240 88	0.874 680 1
		2.768	2.8×10^{-5}	0.18	0.873 9
		2.788	1.6×10^{-5}	0.36	0.875 6
		2.81	0.7×10^{-5}	0.54	0.877
9	0	2.211 05	3.12×10^{-6}		
		2.211 1	3.12×10^{-6}		
		2.209 5	2.9×10^{-6}	1.62	0.939 9
		2.211 15	3.18×10^{-6}	1.572	0.940 151
		2.211 004	3.112×10^{-6}	1.583	0.940 138 2
		2.202	3.5×10^{-6}	1.53	0.939 2
		2.209 9	3.3×10^{-6}	1.56	0.940 04
		2.211 4	3.02×10^{-6}	1.587	0.940 17

Table 8.10(h). $F = 1.5 \times 10^{-5}$au, $|m| = 1, n = 10$.

n_1	n_2	$-E \times 10^3$	Γ	K	Z_1
0	8	7.155	1.5×10^{-4}		
		7.155 4	1.5×10^{-4}		0.121 2
		7.155 33	1.785×10^{-4}	−0.931	0.121 04
		7.155 263 8	$1.792\,0 \times 10^{-4}$	−0.934 35	0.121 156 58
		7.155 263 09	$1.792\,0 \times 10^{-4}$	−0.934 35	0.121 156 616
		7.166	1.67×10^{-4}	−0.84	0.121 12
		7.154	1.70×10^{-4}	−0.95	0.121 146
		7.147	1.59×10^{-4}	−1.0	0.121 09
1	7	6.702	0.90×10^{-4}		
		6.702 2	0.90×10^{-4}		0.237 9
		6.702 059	1.031×10^{-4}	−0.499	0.237 80
		6.702 075 6	$1.035\,6 \times 10^{-4}$	−0.501 96	0.237 923 51
		6.702 074 75	$1.035\,5 \times 10^{-4}$	−0.501 95	0.237 923 536
		6.711	0.881×10^{-4}	−0.41	0.238 0
		6.694	0.879×10^{-4}	−0.57	0.237 80
		6.679	0.73×10^{-4}	−0.71	0.237 6
2	6	6.238	5.0×10^{-5}		
		6.238 2	5.0×10^{-5}		0.350 0
		6.238 09	5.36×10^{-5}	−0.036	0.349 92
		6.238 175 7	$5.384\,5 \times 10^{-5}$	−0.039 999	0.350 039 107
		6.238 174 69	$5.384\,5 \times 10^{-5}$	−0.039 996	0.350 039 113
		6.238 9	3.34×10^{-5}	−0.029	0.349 94
		6.21	3.31×10^{-5}	−0.3	0.349 4
		6.19	2.8×10^{-5}	−0.5	0.348 8
3	5	5.766 0	2.3×10^{-5}		
		5.766 0	2.3×10^{-5}		0.457 3
		5.765 91	2.36×10^{-5}	0.496	0.457 20
		5.766 053 0	$2.377\,5 \times 10^{-5}$	0.491 52	0.457 321 22
		5.766 051 63	$2.377\,5 \times 10^{-5}$	0.491 52	0.457 321 203
		5.755	2.6×10^{-5}	0.40	0.456 9
		5.772	1.7×10^{-5}	0.55	0.457 5
		5.79	0.9×10^{-5}	0.68	0.458 0
4	4	5.289 2	8.01×10^{-6}		
		5.289 2	8.01×10^{-6}		0.559 7
		5.289 0	7.93×10^{-6}	1.154	0.559 6
		5.289 235	$8.002\,6 \times 10^{-6}$	1.148 4	0.559 712 96
		5.289 233 1	$8.002\,5 \times 10^{-6}$	1.148 4	0.559 712 915
		5.279	9.9×10^{-6}	1.05	0.559 2
		5.288 5	7.90×10^{-6}	1.141	0.559 68
		5.292	6.9×10^{-6}	1.18	0.559 84

(Continued)

Table 8.10(h). (*Continued*)

n_1	n_2	$-E \times 10^3$	Γ	K	Z_1
5	3	4.811 3	1.74×10^{-6}		
		4.811 3	1.74×10^{-6}		0.657 3
		4.811 1	1.73×10^{-6}	2.005	0.657 16
		4.811 304	$1.758\,2 \times 10^{-6}$	1.997 6	0.657 280 48
		4.811 301 7	$1.758\,1 \times 10^{-6}$	1.997 7	0.657 280 371
		4.804	2.1×10^{-6}	1.93	0.656 8
		4.810 3	1.80×10^{-6}	1.988	0.657 23
		4.811 25	1.750×10^{-6}	1.997 2	0.657 278
6	2	4.333 93	2.184×10^{-7}		
		4.333 9	2.184×10^{-7}		0.750 1
		4.333 7	2.14×10^{-7}	3.12	0.749 98
		4.333 935	$2.185\,8 \times 10^{-7}$	3.107 3	0.750 102 6
		4.333 931 0	$2.185\,4 \times 10^{-7}$	3.107 4	0.750 102 35
		4.329	2.4×10^{-7}	3.07	0.749 7
		4.333 5	2.21×10^{-7}	3.103	0.750 08
		4.333 82	2.192×10^{-7}	3.106 3	0.750 096
7	1	3.856 779 8	1.311×10^{-8}		
		3.856 8	1.311×10^{-8}		0.838 2
		3.856 4	1.26×10^{-8}	4.58	0.838 05
		3.856 787	$1.312\,4 \times 10^{-8}$	4.564 9	0.838 182 8
		3.856 779 5	$1.311\,2 \times 10^{-8}$	4.565 4	0.838 182 30
		3.853	1.37×10^{-8}	4.55	0.837 8
		3.856 6	1.317×10^{-8}	4.563 5	0.838 173
		3.856 76	$1.312\,2 \times 10^{-8}$	4.565 1	0.838 180 7
8	0	3.379 160 96	2.606×10^{-10}		
		3.379 2	2.606×10^{-10}		0.921 5
		3.378 7	2.4×10^{-10}	6.61	0.921 3
		3.379 19	2.637×10^{-10}	6.562	0.921 490
		3.379 153	$2.581\,6 \times 10^{-10}$	6.572 1	0.921 487 8
		3.376	2.52×10^{-10}	6.59	0.921 1
		3.379 11	2.642×10^{-10}	6.561	0.921 484
		3.379 157	$2.581\,6 \times 10^{-10}$	6.572 1	0.921 488 1

Table 8.10(i). $F = 1.5 \times 10^{-5}$au, $|m| = 1, n = 9$.

n_1	n_2	$-E \times 10^3$	Γ	K	Z_1
0	7	7.739 156 2	1.746×10^{-8}		
					0.125 8
		7.739 5	1.72×10^{-8}	4.504	0.125 72
		7.739 156 8	$1.746 0 \times 10^{-8}$	4.497 2	0.125 830 54
		7.739 156 204	$1.746 0 \times 10^{-8}$	4.497 2	0.125 830 561
		7.735	1.9×10^{-8}	4.46	0.125 69
		7.739 01	1.753×10^{-8}	4.495 8	0.125 829 3
		7.739 13	$1.747 4 \times 10^{-8}$	4.496 9	0.125 830 3
1	6	7.332 737 51	3.104×10^{-9}		
					0.248 1
		7.333 0	3.06×10^{-9}	5.397	0.247 96
		7.332 738 2	$3.104 9 \times 10^{-9}$	5.389 6	0.248 061 40
		7.332 737 516	$3.104 9 \times 10^{-9}$	5.389 6	0.248 061 417
		7.329	3.3×10^{-9}	5.36	0.247 90
		7.332 65	3.112×10^{-9}	5.388 7	0.248 060 0
		7.332 726	$3.105 9 \times 10^{-9}$	5.389 5	0.248 061 2
2	5	6.924 478 531	4.281×10^{-10}		
					0.366 7
		6.924 8	4.22×10^{-10}	6.415	0.366 55
		6.924 479 4	$4.280 8 \times 10^{-10}$	6.407 5	0.366 656 416
		6.924 478 529	$4.280 7 \times 10^{-10}$	6.407 5	0.366 656 420
		6.921	4.5×10^{-10}	6.38	0.366 48
		6.924 42	4.286×10^{-10}	6.406 9	0.366 655 1
		6.924 474	$4.281 3 \times 10^{-10}$	6.407 4	0.366 656 31
3	4	6.514 366 972	4.362×10^{-11}		
					0.481 6
		6.514 6	4.28×10^{-11}	7.584	0.481 48
		6.514 368 0	$4.362 6 \times 10^{-11}$	7.575 0	0.481 579 681
		6.514 366 967	$4.362 5 \times 10^{-11}$	7.575 0	0.481 579 670
		6.512	4.5×10^{-11}	7.56	0.481 39
		6.514 33	4.366×10^{-11}	7.574 7	0.481 578 6
		6.514 365	$4.362 7 \times 10^{-11}$	7.575 0	0.481 579 61
4	3	6.102 417 677	$3.068 2 \times 10^{-12}$		
					0.592 8
		6.102 6	3.00×10^{-12}	8.94	0.592 70
		6.102 419 0	$3.068 4 \times 10^{-12}$	8.926 8	0.592 796 64
		6.102 417 667	$3.068 2 \times 10^{-12}$	8.926 9	0.592 796 609
		6.100	3.16×10^{-12}	8.91	0.592 60
		6.102 40	3.070×10^{-12}	8.926 6	0.592 795 7
		6.102 416 8	$3.068 3 \times 10^{-12}$	8.926 9	0.592 796 57

(*Continued*)

Table 8.10(i). (*Continued*)

n_1	n_2	$-E \times 10^3$	Γ	K	Z_1
5	2	5.688 666 472	$1.340\,1 \times 10^{-13}$		
					0.700 3
		5.688 8	1.30×10^{-13}	10.53	0.700 18
		5.688 668	$1.340\,3 \times 10^{-13}$	10.516	0.700 274 92
		5.688 666 45	$1.340\,1 \times 10^{-13}$	10.516	0.700 274 846
		5.687	1.36×10^{-13}	10.508	0.700 07
		5.688 65	$1.340\,8 \times 10^{-13}$	10.516	0.700 274 1
		5.688 666 1	$1.340\,1 \times 10^{-13}$	10.516	0.700 274 83
6	1	5.273 166 722	$3.027\,3 \times 10^{-15}$		
					0.804 0
		5.273 26	2.90×10^{-15}	12.46	0.803 89
		5.273 170	3.030×10^{-15}	12.433	0.803 985 0
		5.273 166 60	$3.026\,9 \times 10^{-15}$	12.434	0.803 984 840
		5.271	3.021×10^{-15}	12.435	0.803 78
		5.273 156	3.031×10^{-15}	12.433	0.803 984 3
		5.273 166 52	$3.026\,9 \times 10^{-15}$	12.434	0.803 984 835
7	0	4.855 987 480	7.3×10^{-17}		
					0.903 9
		4.855 999	2.1×10^{-17}	14.93	0.903 80
		4.856 000	2.384×10^{-17}	14.877	0.903 901 4
		4.855 984	$2.335\,2 \times 10^{-17}$	14.888	0.903 900 3
		4.854	2.2×10^{-17}	14.91	0.903 69
		4.855 980	2.385×10^{-17}	14.877	0.903 900 1
		4.855 987 38	$2.335\,0 \times 10^{-17}$	14.888	0.903 900 55

Table 8.10(j). $F = 2.0 \times 10^{-5}$au, $|m| = 1, n = 10$.

n_1	n_2	$-E \times 10^3$	Γ	K	Z_1
0	8	7.75	1.9×10^{-3}		
		7.657 1	5.19×10^{-3}	−4.978	0.125 50
		7.656 228 4	$5.224 1 \times 10^{-3}$	−4.986 8	0.125 639 90
		7.656 227 35	$5.224 1 \times 10^{-3}$	−4.986 8	0.125 639 953
		7.663	5.10×10^{-3}	−4.93	0.125 55
		7.656 4	5.223×10^{-3}	−4.985	0.125 641 5
		7.656 27	$5.224 2 \times 10^{-3}$	−4.986 5	0.125 640 3
1	7	7.20	1.6×10^{-3}		
		7.172 9	2.69×10^{-3}	−3.982	0.247 26
		7.172 272 6	$2.706 6 \times 10^{-3}$	−3.990 0	0.247 397 55
		7.172 271 53	$2.706 6 \times 10^{-3}$	−3.990 0	0.247 397 591
		7.180	2.64×10^{-3}	−3.93	0.247 37
		7.172 7	$2.706 4 \times 10^{-3}$	−3.987	0.247 404
		7.172 37	$2.706 8 \times 10^{-3}$	−3.989 2	0.247 399 2
2	6	6.67	0.97×10^{-3}		
		6.657 2	1.43×10^{-3}	−3.076	0.364 42
		6.656 856	$1.442 0 \times 10^{-3}$	−3.082 3	0.364 549 98
		6.656 854 5	$1.442 0 \times 10^{-3}$	−3.082 3	0.364 550 003
		6.665	1.41×10^{-3}	−3.02	0.364 60
		6.657 6	$1.442 3 \times 10^{-3}$	−3.077	0.364 567
		6.657 1	$1.442 0 \times 10^{-3}$	−3.080 8	0.364 555
3	5	6.11	5.7×10^{-4}		
		6.107 82	7.67×10^{-4}	−2.257	0.476 17
		6.107 796	$7.706 9 \times 10^{-4}$	−2.262 9	0.476 301 221
		6.107 794 1	$7.706 9 \times 10^{-4}$	−2.262 9	0.476 301 211
		6.117	7.5×10^{-4}	−2.19	0.476 46
		6.109 0	$7.707 3 \times 10^{-4}$	−2.254	0.476 34
		6.107 85	7.69×10^{-4}	−2.262 5	0.476 303
4	4	5.524	3.1×10^{-4}		
		5.523 0	3.95×10^{-4}	−1.518	0.581 65
		5.523 303	$3.969 1 \times 10^{-4}$	−1.523 2	0.581 792 18
		5.523 300 2	$3.969 1 \times 10^{-4}$	−1.523 2	0.581 792 109
		5.534	3.84×10^{-4}	−1.43	0.582 1
		5.524 4	3.94×10^{-4}	−1.515	0.581 84
		5.521	3.89×10^{-4}	−1.54	0.581 72
5	3	4.904	1.6×10^{-4}		
		4.903 4	1.82×10^{-4}	−0.823	0.680 0
		4.903 994	$1.837 5 \times 10^{-4}$	−0.828 55	0.680 207 0
		4.903 989 0	$1.837 4 \times 10^{-4}$	−0.828 52	0.680 206 80

(Continued)

Table 8.10(j). (*Continued*)

n_1	n_2	$-E \times 10^3$	Γ	K	Z_1
		4.92	1.68×10^{-4}	−0.73	0.680 6
		4.901	1.72×10^{-4}	−0.85	0.680 1
		4.891	1.57×10^{-4}	−0.93	0.679 6
6	2	4.254	6.3×10^{-5}		
		4.253 4	6.71×10^{-5}	−0.092	0.770 8
		4.254 395	$6.788\,8 \times 10^{-5}$	−0.099 98	0.770 952 7
		4.254 384 6	$6.787\,9 \times 10^{-5}$	−0.099 893	0.770 952 14
		4.257	4.6×10^{-5}	−0.07	0.770 957
		4.23	4.4×10^{-5}	−0.3	0.769
		4.20	3.6×10^{-5}	−0.5	0.768
7	1	3.585 1	1.56×10^{-5}		
		3.583 9	1.53×10^{-5}	0.84	0.853 8
		3.585 17	$1.566\,5 \times 10^{-5}$	0.824 0	0.853 969
		3.585 143	$1.565\,4 \times 10^{-5}$	0.824 42	0.853 967 8
		3.571	1.9×10^{-5}	0.73	0.852 9
		3.586 0	1.4×10^{-5}	0.831	0.854 02
		3.594	1.1×10^{-5}	0.90	0.854 5
8	0	2.913 82	1.154×10^{-6}		
		2.912	1.06×10^{-6}	2.32	0.929 8
		2.913 91	1.17×10^{-6}	2.269	0.930 062
		2.913 780	1.144×10^{-6}	2.279 4	0.930 051 8
		2.905	1.24×10^{-6}	2.252	0.929 3
		2.912 8	1.20×10^{-6}	2.259	0.929 98
		2.913 62	1.146×10^{-6}	2.278 0	0.930 040

Table 8.10(k). $F = 2.0 \times 10^{-5}$au, $|m| = 1, n = 9$.

n_1	n_2	$-E \times 10^3$	Γ	K	Z_1
0	7	8.385 4	3.72×10^{-5}		
		8.385 72	3.75×10^{-5}	0.368	0.131 11
		8.385 631 7	$3.769 7 \times 10^{-5}$	0.364 17	0.131 242 25
		8.385 630 62	$3.769 7 \times 10^{-5}$	0.364 18	0.131 242 286
		8.37	3.85×10^{-5}	0.28	0.131 0
		8.398	2.2×10^{-5}	0.45	0.131 33
		8.42	1.1×10^{-5}	0.6	0.131 5
1	6	7.830 3	1.63×10^{-5}		
		7.830 41	1.66×10^{-5}	0.872	0.257 40
		7.830 359 5	$1.675 9 \times 10^{-5}$	0.868 04	0.257 536 50
		7.830 358 18	$1.675 9 \times 10^{-5}$	0.868 05	0.257 536 523
		7.816	2.1×10^{-5}	0.76	0.257 2
		7.831 2	1.5×10^{-5}	0.874	0.257 549
		7.840	1.2×10^{-5}	0.94	0.257 7
2	5	7.271 4	5.759×10^{-6}		
		7.271 42	5.73×10^{-6}	1.494	0.378 67
		7.271 400	$5.788 3 \times 10^{-6}$	1.488 5	0.378 803 008
		7.271 398 0	$5.788 3 \times 10^{-6}$	1.488 5	0.378 803 010
		7.259	7.1×10^{-6}	1.40	0.378 4
		7.269 9	5.90×10^{-6}	1.477	0.378 77
		7.272 5	5.5×10^{-6}	1.497	0.378 83
3	4	6.710 94	1.410×10^{-6}		
		6.710 934	1.395×10^{-6}	2.273	0.494 90
		6.710 940	$1.413 2 \times 10^{-6}$	2.266 0	0.495 030 28
		6.710 937 6	$1.413 1 \times 10^{-6}$	2.266 0	0.495 030 251
		6.702	1.63×10^{-6}	2.21	0.494 6
		6.709 9	1.45×10^{-6}	2.258	0.495 00
		6.710 75	1.415×10^{-6}	2.264 6	0.495 025
4	3	6.149 670	2.257×10^{-7}		
		6.149 62	2.22×10^{-7}	3.249	0.606 09
		6.149 673	$2.256 1 \times 10^{-7}$	3.239 9	0.606 219 37
		6.149 670 4	$2.256 0 \times 10^{-7}$	3.239 9	0.606 219 300
		6.143	2.5×10^{-7}	3.20	0.605 8
		6.149 2	2.28×10^{-7}	3.236	0.606 201
		6.149 55	2.262×10^{-7}	3.239 0	0.606 215
5	2	5.587 228 4	$2.156 0 \times 10^{-8}$		
		5.587 11	2.10×10^{-8}	4.47	0.712 22
		5.587 232	$2.155 6 \times 10^{-8}$	4.460 0	0.712 346 51
		5.587 228 3	$2.155 2 \times 10^{-8}$	4.460 1	0.712 346 35
		5.582	2.3×10^{-8}	4.43	0.712 0

(*Continued*)

Table 8.10(k). (*Continued*)

n_1	n_2	$-E \times 10^3$	Γ	K	Z_1
		5.587 0	2.164×10^{-8}	4.458 5	0.712 338
		5.587 19	2.157×10^{-8}	4.459 8	0.712 344 8
6	1	5.023 144 12	$1.043 6 \times 10^{-9}$		
		5.022 9	1.00×10^{-9}	6.03	0.813 2
		5.023 152	$1.044 3 \times 10^{-9}$	6.013 5	0.813 373 6
		5.023 143 8	$1.043 4 \times 10^{-9}$	6.013 9	0.813 373 17
		5.019	1.07×10^{-9}	6.002	0.813 0
		5.023 06	1.046×10^{-9}	6.012 8	0.813 369
		5.023 136	$1.043 6 \times 10^{-9}$	6.013 9	0.813 372 8
7	0	4.457 214 621	$1.720 8 \times 10^{-11}$		
		4.456 8	1.56×10^{-11}	8.15	0.909 1
		4.457 24	1.742×10^{-11}	8.096 2	0.909 267
		4.457 206	$1.705 4 \times 10^{-11}$	8.106 6	0.909 265 1
		4.454	1.65×10^{-11}	8.12	0.908 9
		4.457 18	1.744×10^{-11}	8.095 6	0.909 263
		4.457 213	$1.705 3 \times 10^{-11}$	8.106 7	0.909 265 5

Table 8.10(l). $F = 2.0 \times 10^{-5}$au, $|m| = 1, n = 8$.

n_1	n_2	$-E \times 10^3$	Γ	K	Z_1
0	6	9.375 532 526	$3.235\,8 \times 10^{-11}$		
		9.376 1	3.18×10^{-11}	7.890	0.138 38
		9.375 533 5	$3.235\,9 \times 10^{-11}$	7.881 1	0.138 495 94
		9.375 532 525	$3.235\,8 \times 10^{-11}$	7.881 1	0.138 495 970
		9.372	3.4×10^{-11}	7.86	0.138 35
		9.375 49	3.238×10^{-11}	7.880 8	0.138 495 7
		9.375 530	$3.236\,0 \times 10^{-11}$	7.881 1	0.138 495 95
1	5	8.895 826 102	$3.621\,2 \times 10^{-12}$		
		8.896 4	3.56×10^{-12}	9.01	0.273 10
		8.895 827 1	$3.621\,3 \times 10^{-12}$	8.998 0	0.273 210 896
		8.895 826 100	$3.621\,3 \times 10^{-12}$	8.998 0	0.273 210 910 4
		8.893	3.8×10^{-12}	8.98	0.273 05
		8.895 80	3.623×10^{-12}	8.997 8	0.273 210 5
		8.895 825 0	$3.621\,3 \times 10^{-12}$	8.998 0	0.273 210 89
2	4	8.413 807 725	$3.005\,4 \times 10^{-13}$		
		8.414 3	2.94×10^{-13}	10.274	0.404 00
		8.413 808 9	$3.005\,3 \times 10^{-13}$	10.264	0.404 104 611 8
		8.413 807 720	$3.005\,2 \times 10^{-13}$	10.264	0.404 104 611 5
		8.411	3.10×10^{-13}	10.249	0.403 93
		8.413 79	3.007×10^{-13}	10.264	0.404 104 1
		8.413 807 1	$3.005\,3 \times 10^{-13}$	10.264	0.404 104 599
3	3	7.929 521 338	$1.664\,8 \times 10^{-14}$		
		7.930 0	1.68×10^{-14}	11.73	0.531 03
		7.929 523	$1.725\,5 \times 10^{-14}$	11.713	0.531 138 70
		7.929 521 328	$1.725\,4 \times 10^{-14}$	11.713	0.531 138 676
		7.927	1.76×10^{-14}	11.703	0.530 95
		7.929 51	$1.726\,0 \times 10^{-14}$	11.713	0.531 138 2
		7.929 521 0	$1.725\,4 \times 10^{-14}$	11.713	0.531 138 667
4	2	7.443 022 725	5.620×10^{-16}		
		7.443 4	6.0×10^{-16}	13.42	0.654 18
		7.443 025	6.165×10^{-16}	13.399	0.654 277 04
		7.443 022 70	$6.163\,4 \times 10^{-16}$	13.399	0.654 276 985
		7.441	6.24×10^{-16}	13.394	0.654 08
		7.443 01	6.166×10^{-16}	13.399	0.654 276 6
		7.443 022 5	$6.163\,4 \times 10^{-16}$	13.399	0.654 276 980
5	1	6.954 379 305	1.20×10^{-17}		
		6.954 7	1.09×10^{-17}	15.44	0.773 39
		6.954 383	$1.141\,5 \times 10^{-17}$	15.413	0.773 486 31
		6.954 379 18	$1.140\,5 \times 10^{-17}$	15.413	0.773 486 152

(Continued)

Table 8.10(1). (*Continued*)

n_1	n_2	$-E \times 10^3$	Γ	K	Z_1
		6.952	1.133×10^{-17}	15.417	0.773 28
		6.954 371	$1.141\,8 \times 10^{-17}$	15.413	0.773 485 8
		6.954 379 20	$1.140\,5 \times 10^{-17}$	15.413	0.773 486 154
6	0	6.463 670 073	$< 10^{-17}$		
		6.463 9	6.6×10^{-20}	18.01	0.888 64
		6.463 68	7.369×10^{-20}	17.953	0.888 736 7
		6.463 666	$7.219\,7 \times 10^{-20}$	17.964	0.888 735 8
		6.462	6.8×10^{-20}	17.99	0.888 53
		6.463 664	7.372×10^{-20}	17.953	0.888 735 7
		6.463 670 02	$7.219\,2 \times 10^{-20}$	17.964	0.888 736 00

References

Abramowitz, M., and Stegun, I. A., Editors, 1965, *Handbook of Mathematical Functions.* National Bureau of Standards, Applied Mathematics Series **55**. Fourth Printing, with corrections, Washington D. C.

Alexander, M. H., 1969, *Phys. Rev.* **178**, 34–40.

Alliluev, S. P., and Malkin, I. A., 1974, *Zh. Eksp. Teor. Fiz.* **66**, 1283–1294. English translation: *Sov. Phys. JETP* **39**, 627–632 (1974).

Andersson, N., Araújo, M. E., and Schutz, B. F., 1993, *Class. Quantum Grav.* **10**, 735–755.

Athavan, N., Fröman, P. O., Fröman, N., and Lakshmanan, M., 2001a, *J. Math. Phys.* **42**, 5051–5076.

Athavan, N., Lakshmanan, M., and Fröman, N., 2001b, *J. Math. Phys.* **42**, 5077–5095.

Athavan, N., Lakshmanan, M., and Fröman, N., 2001c, *J. Math. Phys.* **42**, 5096–5115.

Bailey, D. S., Hiskes, J. R., and Riviere, A. C., 1965, *Nuclear Fusion* **5**, 41–46.

Basu, K., 1934, *Bulletin of the Calcutta Mathematical Society* **26**, Nos. 3–4, 78.

Bayfield, J. E., 1979, *Phys. Rep.* **51**, 317–391.

Bekenstein, J. D., and Krieger, J. B., 1969, *Phys. Rev.* **188**, 130–139.

Bethe, H. A., and Salpeter, E. E., 1957, *Quantum Mechanics of One- and Two-electron Atoms.* Springer-Verlag, Berlin, Göttingen and Heidelberg.

Born, M., and Jordan, P., 1925, *Z. Physik* **34**, 858–888.

Born, M., Heisenberg, W., and Jordan, P., 1926, *Z. Physik* **35**, 557–615.

Byrd, P. F., and Friedman, M. D., 1971,*Handbook of Elliptic Integrals for Engineers and Scientists*, Second edition, Revised. Springer-Verlag, Berlin, Heidelberg and New York.

Damburg, R. J., and Kolosov, V. V., 1976a, *J. Phys. B: Atom. Molec. Phys.* **9**, 3149–3157.

Damburg, R. J., and Kolosov, V. V., 1976b, *Fifth Int. Conf. on Atomic Physics*, Berkeley, California, Abstracts pages 202–203.

Damburg, R. J., and Kolosov, V. V., 1977, *Phys. Lett.* **61A**, 233–234.

Damburg, R. J., and Kolosov, V. V., 1978a, *Opt. Commun.* **24**, 211–212.

Damburg, R. J., and Kolosov, V. V., 1978b, *J. Phys. B: Atom. Molec. Phys.* **11**, 1921–1930.

Damburg, R. J., and Kolosov, V. V., 1979, *J. Phys. B: Atom. Molec. Phys.* **12**, 2637–2643.

Damburg, R. J., and Kolosov, V. V., 1980, Article on pp. 741–750 in *Proceedings of the XIth International Conference on the Physics of Electronic and Atomic Collisions*, Kyoto, 29 August–4 September 1979, edited by N. Oda and K. Takayanagi. North-Holland Publishing Company, Amsterdam, New York, Oxford.

Damburg, R. J., and Kolosov, V. V., 1981, *J. Phys. B: Atom. Molec. Phys.* **14**, 829–834.

Damburg, R. J., and Kolosov, V. V, 1982, Article on pp. 31–71 in *Rydberg States of Atoms and Molecules*, edited by R. F. Stebbings and B. F. Dunning. Cambridge University Press.

Dammert, Ö., and Fröman, P. O., 1980, *J. Math. Phys.* **21**, 1683–1687

Dirac, P. A. M., 1925, *Proc. Roy. Soc.* **109**, 642–653.

Drukarev, G. F., 1978, *Zh. Eksp. Teor. Fiz.* **75**, 473–483. English translation: *Sov. Phys. JETP* **48**, 237–242 (1978).

Drukarev, G. F., 1982, *Zh. Eksp. Teor. Fiz.* **82**, 1388–1392. English translation: *Sov. Phys. JETP* **55**, 806–808 (1982).

Drukarev, G., Fröman, N., and Fröman, P. O., 1979, *J. Phys. A: Math. Gen.* **12**, 171–186.

Epstein, P. S., 1926, *Phys. Rev.* **28**, 695–710.

Farrelly, D., and Reinhardt, W. P., 1983, *J. Phys. B: Atom. Molec. Phys.* **16**, 2103–2117.

Froelich, P., and Brändas, E., 1975, *Phys. Rev.* **A12**, 1–5.

Fröman, N., 1966a, *Ark. Fys.* **31**, 381–408.

Fröman, N., 1966b, *Ark. Fys.* **32**, 79–97.

Fröman, N., 1966c, *Ark. Fys.* **32**, 541–548.

Fröman, N, 1970, *Ann. Phys. (N.Y.)* **61**, 451–464.

Fröman, N., 1978, *Phys. Rev.* **A17**, 493–504.

Fröman, N., 1979, *J. Phys. A: Math. Gen.* **12**, 2355–2371.

Fröman, N., 1980, Article on pp. 1–44 in *Semiclassical Methods in Molecular Scattering and Spectroscopy*, edited by M. S. Child. D. Reidel Publishing Company, Dordrecht, Boston, London.

Fröman, N., and Fröman, P. O., 1965, *JWKB Approximation, Contributions to the Theory.* North-Holland Publishing Company, Amsterdam. Russian translation: MIR, Moscow 1967.

Fröman, N., and Fröman, P. O., 1978a, *J. Math. Phys.* **19**, 1823–1829.

Fröman, N., and Fröman, P. O., 1978b, *J. Math. Phys.* **19**, 1830–1837.

Fröman, N., and Fröman, P. O., 1978c, *Ann. Phys. (N. Y.)* **115**, 269–275.

Fröman, N., and Fröman, P. O., 1984, Phase-integral calculation with very high accuracy of the Stark effect in a hydrogen atom. Colloque du 10 au 15 septem-ber 1984, CIRM (Luminy), sur *Méthodes Semi-Classiques en Mécanique Quantique*, p. 45 (edited by B. Helffer and D. Robert). Publications de l'Université de Nantes, Institut de Mathématiques et d'Informatiques, 2 rue de la Houssinière, 44072 Nantes CEDEX, France.

Fröman, N., and Fröman, P. O., 1985, *On the Historyy of the So-Called WKB Method from 1817 to 1926.* Proceedings of the Niels Bohr Centennial Conference, Copenhagen 25–28 March 1985 on Semiclassical Descriptions of Atomic and Nuclear Collisions, pp. 1–7, edited by J. Bang and J. de Boer. North-Holland Publishing Company, Amsterdam, Oxford, New York and Tokyo.

Fröman, N., and Fröman, P. O., 1996, *Phase-Integral Method Allowing Nearlying Transition Points*, with adjoined papers by A. Dzieciol, N. Fröman, P. O. Fröman, A. Hökback, S. Linnæus, B. Lundborg, and E. Walles. Springer Tracts in Natural Philosophy **40**, edited by C. Truesdell. Springer-Verlag, New York, Berlin and Heidelberg.

Fröman, N., and Fröman, P. O., 2002, *Physical Problems Solved by the Phase-Integral Method.* Cambridge University Press. Paperback 2005.

Fröman, N., and Mrazek, W., 1977, *J. Phys. A: Math. Gen.* **10**, 1287–1295

Fröman, N., Fröman, P. O., and Larsson, K., 1994, *Phil. Trans. Roy. Soc. Lond.* **A347**, 1–22.

Fröman, N., Fröman, P. O., and Lundborg, B., 1988, *Math. Proc. Camb. Phil. Soc.* **104**, 153–179.

Fröman, N., Fröman, P. O., and Lundborg, B., 1996. This is the adjoined paper Chapter 5 in Fröman and Fröman (1996).

Fröman, N., Fröman, P. O., Andersson, N., and Hökback, A., 1992, *Phys. Rev.* **D45**, 2609–2616.

Gallagher, T. F., 1988, *Rep. Prog. Phys.* **51**, 143–188.

Gallagher, T. F., 1994, *Rydberg Atoms.* Cambridge Monographs on Atomic, Molecular and Chemical Physics **3**, Cambridge University Press.

Gallas, J. A. C., Walther, H., and Werner, E., 1982a, *Phys. Rev.* **A26**, 1775–1778.

Gallas, J. A. C., Walther, H., and Werner, E., 1982b, *Phys. Rev. Lett.* **49**, 867–870.

Gallas, J. A. C., Leuchs, G., Walther, H., and Figger, H., 1985, *Advances in Atomic and Molecular Physics* **20**, 413–466.

Guschina, N. A., and Nikulin, V. K., 1975, *Chem. Phys.* **10**, 23–31.

Harmin, D. A., 1981, *Phys. Rev.* **A24**, 2491–2512.

Hehenberger, M., McIntosh, H. V., and Brändas, E., 1974, *Phys. Rev.* **A10**, 1494–1506.

Heisenberg, W., 1925, *Z. f. Physik* **33**, 879–893.

Herrick, D. R., 1976, *J. Chem. Phys.* **65**, 3529–3535.

Hirschfelder, J. O., and Curtiss, L. A., 1971, *J. Chem. Phys.* **55**, 1395–1402.

Jeffreys, H., 1925, *Proc. Lond. Math. Soc.* (Second Series) **23**, 428–436.

Koch, P. M., 1981, Article on pp. 181–207 in *Atomic Physics* **7**, edited by D. Kleppner and F. M. Pipkin. Plenum Press, New York and London.

Kolosov, V. V., 1983, *J. Phys. B: Atom. Molec. Phys.* **16**, 25–31.

Kolosov, V. V., 1987, *J. Phys. B: Atom. Molec. Phys.* **20**, 2359–2367.

Korsch, H. J., and Möhlenkamp, R., 1983, *Z. Phys. A — Atoms and Nuclei* **314**, 267–273.

Krylov, N. S., and Fock, V. A., 1947, *Zh. Eksp. Teor. Fiz.* **17**, 93–107.

Lanczos, C., 1930a, *Die Naturwissenschaften* **18**, 329–330.

Lanczos, C., 1930b, *Z. Physik* **62**, 518–544.

Lanczos, C., 1930c, *Z. Physik* **65**, 431–455.

Lanczos, C., 1931, *Z. Physik* **68**, 204–232.

Lisitsa, V. S., 1987, *Usp. Fiz. Nauk* **153**, 379–421. English translation: *Sov. Phys. Usp.* **30**, 927–951 (1987).

Luc-Koenig, E., and Bachelier, A., 1980a, *J. Phys. B: Atom. Molec. Phys.* **13**, 1743–1767.

Luc-Koenig, E., and Bachelier, A., 1980b, *J. Phys. B: Atom. Molec. Phys.* **13**, 1769–1790.

Oppenheimer, J. R., 1928, *Phys. Rev.* **31**, 66–81.

Pauli, W., 1926, *Z. Physik* **36**, 336–363.

Paulsson, R., Karlsson, F., and LeRoy, R. J., 1983, *J. Chem. Phys.* **79**, 4346–4354.

Rice, M. H., and Good, R. H., Jr., 1962, *J. Opt. Soc. Am.* **52**, 239–246.

Rojas, C., and Villalba, V. M., 2007, *Phys. Rev.* **D75**, 063518-1.

Ryde, N., 1976, *Atoms and Molecules in Electric Fields.* Almqvist and Wiksell International, Stockholm.

Schrödinger, E., 1926, *Annalen der Physik (Vierte Folge)* **80**, 437–490.

Silverstone, H. J., and Koch, P. M., 1979, *J. Phys B: Atom. Molec. Phys.* **12**, L537–L541.

Van Vleck, J. H., 1926, *Proc. N.A.S.* **12**, 662–670.

Waller, I., 1926, *Z. Physik* **38**, 635–646.

Wentzel, G., 1926, *Z. Physik* **38**, 518–529.

Yamabe, T., Tachibana, A., and Silverstone, H. J., 1977, *Phys. Rev.* **A16**, 877–890.

Name Index

Subject Index